J. L Pulvermacher

Galvanic Electricity

Its Pre-Eminent Power and Effects in Preserving and Restoring Health

Made Plain and Useful

J. L Pulvermacher

Galvanic Electricity
Its Pre-Eminent Power and Effects in Preserving and Restoring Health Made Plain and Useful

ISBN/EAN: 9783337864217

Printed in Europe, USA, Canada, Australia, Japan

Cover: Foto ©berggeist007 / pixelio.de

More available books at **www.hansebooks.com**

GALVANIC ELECTRICITY;

ITS

PRE-EMINENT POWER AND EFFECTS

IN

PRESERVING AND RESTORING HEALTH

MADE

PLAIN AND USEFUL.

By J. L. PULVERMACHER.

LONDON:
GALVANIC ESTABLISHMENT, 194 REGENT STREET.
(Opposite Conduit Street.)

1875.

Dedicated

TO THE FRIENDS OF PROGRESS

AND

ALL WHO LOVE COMMON SENSE AND FAIR PLAY.

PREFACE.

THE object of this little work is to meet a long-felt want for a concise Treatise "On Electricity as a Curative" accessible to every understanding. It is intended to furnish those who have but little leisure for a profound study of the electrical sciences, with the means to initiate themselves in some of the *rationale* of the powers and effects of Electricity, and in a manner as simple as are the experiments afforded by my Flexible Galvanic Chain-Bands and Batteries.

The reader will be enabled to judge for himself; become familiar with the why and wherefore of the curative effects of Galvanic Electricity in general, and of Pulvermacher's Medico-Galvanic System in particular. He will also understand why that system has acquired the honourable position it now holds in standard medical works, and the justice of the encomiums bestowed on it by the Scientific Press in general.

To achieve this object is imperative, as frequent attempts have been made by unscrupulous and ignorant speculators to trade upon the pre-eminent reputation of Electricity as a curative agent, and impose upon the credulity of those who have but little acquaintance with physical science.

Having always based my labours upon the recognised sciences, no new theories are advanced, being content to confine myself solely to established scientific facts. Fully aware that in this little work the subject is very far from being exhausted, I hope to do more ample justice to its importance in a Treatise* upon which I am now engaged, and will shortly place before the public. If by these efforts I succeed in diffusing a general knowledge of Electricity, for which my Flexible Chain-Bands and Batteries are so serviceable an auxiliary, I shall not have laboured in vain.

J. L. PULVERMACHER.

194, REGENT STREET, LONDON, W.

March, 1875.

* "GENERALISED MEDICAL ELECTRICITY: How, When, and by What Means best to Apply it Practically and Rationally." Profusely Illustrated.

TABLE OF CONTENTS.

36. Variety of ailments successfully treated by Electricity, recorded from various *bonâ fide* sources.
37. Modus operandi of the physical, chemical, and physiological action of Electricity.
38. Initiatory causes of Electricity promoting circulation, etc.
39. Difference of mode of action by drugs and that of Electricity.
40. Endosmosis and Exosmosis promoted by Electricity.
41. Transport of matter by Electric Currents.
42. The healing properties vary with the varied application.
43. Intricate cases best treated by a scientific medical electrician.
44. Haphazard application frequently successful—always harmless.
45. Illustrative instance of different modes of application curing ailments of an opposite nature.
46. Electricity the greatest modifier of nervous action.
47. Intermittent and continuous currents; their modus operandi.
48. Prolonged application without intervention of conducting wires.
49. Congenial workings of prolonged mild continuous currents.
50. Striking instances, physically, of the peculiar action of prolonged weak currents.
51. Identity of physical Electricity with that in the animal organisation.
52. Nervo-electric currents.
53. Physiological experiments by Brown-Séquard.
54. Nervo-electric currents, when intercepted, cause inflammation—experiment of Claude Bernard.
55. Animal Battery, formed by means of frogs' legs—experiment by Humboldt.
56. Electro-generating organ of Electric Fish producing shocks followed by exhaustion.
57. Nervous exhaustion in man by dispersion of Vital Electricity analogous to that in the Electric Fish.
58. Knowledge is power, proved by progress in the healing art.
59. Origin of Electricity in the system: discovery of Dr. Kincke, of Berlin.
60. Thermo-Electricity, its influence on Vital Electricity.
61. Harmless nature of Electricity—the promoter of its adoption.
62. Electricity the most efficient preventive.
63. Present and future *rôle* of the usefulness of Electricity.
64. Historic sketch of Pulvermacher's inventions in Telegraphy—Electro-motive engines preceding those of his Medico-Galvanic Appliances.
65. Medical Electricity, its mercenary abuse.

GALVANIC ELECTRICITY:

ITS

PRE-EMINENT POWERS AND EFFECTS

MADE

PLAIN AND USEFUL.

ELECTRICITY, ATMOSPHERIC: ITS MAJESTIC PHENOMENA.

1. THERE is nothing in Nature more calculated to impress us with a sense of the majesty and greatness of the Creator than the phenomena of the thunderstorm. Every one is acquainted with the atmospheric changes which accompany it, and has observed the effects it produces on living beings. The oppressiveness which is felt before and during the storm, is followed, when the thunder clouds have exhausted themselves, by a refreshing and reviving influence, which is partly due to the ozone generated by their agency. In olden times, the mighty explosions following the flashes were regarded as the wrathful voice of Heaven incensed at the wickedness of man. During long subsequent ages the causes and effects of the thunderstorm were not understood, and the utility of Electricity was unknown. The spread of education and the progress of science have swept away the superstitious idea which had so

long prevailed in regard to these storms. The power once so dreaded as a tyrannical master has now become, through the aid of science, a most willing and useful servant. It does our errands by sea and land, is capable of driving our machinery, and is a healer of our bodily ailments. We are quite ignorant of the nature of Electricity, or the electric fluid, as it is sometimes called; and although it is everywhere present, we are not aware of its existence until it is awakened from its latent state, that is, excited to action.

2. This peculiar and invisible agency is one of those mysterious powers of Nature only known to us by its effects. Our first acquaintance with it seems to have arisen out of a curious but simple fact, noticed fully 600 years before the Christian era. Thales of Miletus, a celebrated Greek philosopher, observed as a remarkable property of amber, its power of attracting light particles of matter, on being subjected to a peculiar kind of excitation by friction. With this he is said to have been so struck, that he imagined the amber to be endowed with a species of animation. The attractive property developed in amber by the process of friction may be considered as the source of the name bestowed on this power, the Greek word expressive of amber being *electron*, in Latin *electrum*; the unknown principle or element with which Thales supposed it to be animated, has therefore been named ELECTRICITY. As our knowledge of such phenomena advanced, and other substances were observed to possess similar properties, they were termed *electrics*. There are many electrics which are easily procurable, and which serve well for the purpose of experiment. A stick of sealing-wax rubbed with flannel, or a rod of glass rubbed with silk,

exhibits this property, and becomes, as it is called, electrically "excited." If one of these excited bodies be held near the cheek, a sensation will be felt similar to that produced by drawing a cobweb over the face. In the dark, these excited bodies are faintly luminous, and if the finger be held near them, a small spark will dart to it, accompanied with a slight crackling sound.

ELECTRICITY, FRICTIONAL: ITS IDENTITY WITH ATMOSPHERIC.

3. That the power observed in the amber when excited by friction, with which the philosophers of long past ages amused themselves, and the lightning flash which rends the oak and shatters the rock and destroys life and property, were one and the same natural agent, remained unknown until Benjamin Franklin, in the year 1749, first explained the phenomena of thunder gusts, and the Aurora Borealis, upon electrical principles. In the same year he conceived the grand idea of ascertaining the truth of his doctrine, by actually drawing down the lightning, by means of sharp-pointed iron rods raised into the region of the clouds. But it was not until the summer of 1752 that he was enabled by experiment to complete his unparalleled discovery. He prepared a kite, by fastening two cross sticks to a silk handkerchief, which would not suffer so much from the rain as paper. To the upright stick was affixed an iron point. The string was as usual of hemp, except the lower end, which was silk, and where the hempen string terminated a key was fastened. With this apparatus, on the appearance of an approaching thunder-gust, he went on to the common of Philadelphia, accompanied by his son. Repeated sparks were drawn from the key, and many of

the experiments made which are now usually performed with an electrifying machine. This showed to his own satisfaction, and that of all scientific men, the identity of lightning and electricity; and laid the foundation of the use and adoption of lightning conductors, with which we protect our public buildings, houses, and ships, against the ravages of the thunderstorm.

4. About thirty-nine years after this event, Professor Galvani, who held the anatomical chair at the University of Bologna, made, by accident, another wonderful discovery in electrical science. The Professor being one day engaged in experimenting on the frogs which were being made into a soup for his then delicate wife, laid one, prepared after the manner now styled the preparation of Galvani, on a table at no great distance from an electrical machine. One of his assistants, his nephew, Camillo Galvani, having accidentally or designedly touched the crural nerves of this frog with a scalpel whilst the electrical machine was being put in motion, noticed that strong contraction of the hind legs ensued. Madame Galvani, who was present, directed her attention to the circumstance, and further observed, that the convulsive movements invariably occurred at the instant sparks were being drawn from the conductor of the electric machine. Of this fact she informed her husband, who speedily convinced himself of the accuracy of her observation, and immediately commenced a laborious examination of the causes of the phenomena. After repeating and varying the experiment in the most elaborate manner, he announced to the scientific world his conviction of the discovery of an "animal electricity," properly so-called, or, in other words, *that all animals are endowed with an inherent*

constitutional electricity, secreted by the brain and distributed through the nervous system, the principal reservoirs being the muscles.

This announcement was received with enthusiasm throughout Italy and elsewhere, and gave rise to the well-known word Galvanism. Galvani bestowed the name of Animal Electricity on the power which caused spontaneous convulsions in the limbs of a frog when a divided nerve was connected with a muscle by means of a metallic conductor.

ELECTRICITY PRODUCED BY THE VOLTAIC PILE.

5. Volta, the Professor of Natural Philosophy in the University of Pavia, observing that the effects on the frog were far greater when the connecting medium consisted of two different kinds of metal, inferred from this that the principle of excitation existed in the metals and not in the nerves of the animal; and he assumed that by their contact there was developed a small quantity of the electrical fluid, which, being transmitted through the organs of the frog, caused the convulsive movements. These discoveries Volta communicated to the Royal Society of London, in 1793, and in the following year he was awarded the Gold Medal of that Society.

6. Repeated experiments, followed up during seven years, led Volta at length to the invention of what is called a Voltaic Battery. This apparatus, which he named a *corona*, was superseded by one formed on the same principle—namely, successive alternations of heterogeneous metallic plates, which is now known as the Voltaic Pile. This pile Volta invented in 1800, and shortly afterwards the decomposition of water by its means was discovered by Messrs. Nicholson and

Carlisle, and the decomposition of saline chemical compounds by Humphrey Davy. Volta's pile, or electromotive apparatus, as he terms it, which has immortalised his name, was thus constructed: He placed, for example, one pair of metals (zinc and copper) upon another, separating them with a piece of cloth or cardboard, moistened with a saline fluid or water, thereby producing one of the most wonderful combinations for generating electricity ever contemplated; hence the name Voltaism. This Voltaic source of Electricity was the prompting cause of a variety of experiments, by means of which many important electrical discoveries have been made; not the least among them is the method by which the telegraph does its wonderful office. Voltaism has also taken the place of frictional electricity in therapeutics, in consequence of its electro-chemical properties being endowed with superior healing powers. But there was one grave difficulty which greatly interfered with the adoption of the Voltaic pile as a curative agent, and that was its inconstant current and impracticability, which caused it to be abandoned by medical men, so soon as Michael Faraday introduced his magneto-electric machines. These produced only an intermittent current, and consequently their application was always accompanied with muscular contractions or shocks, followed by little or no chemical effect, as was the case with frictional electricity. Although Faraday's invention was a more convenient and ready means for practical medical use than the Voltaic pile, the want of an improved method for generating *continuous currents of electricity* was strongly felt. To supply this want the research and inventive capabilities of scientific electricians throughout the world were enlisted, and resulted in the discovery of a variety of Voltaic combinations or batteries by Wollaston, Cruickshank, Daniell, Bunsen, and others; indeed, the list of the scientific men who, for the benefit of their fellow-creatures, prosecuted these inquiries, is a long and honourable one.

7. Medical Electricity was then regarded as a special science, including Electro-Physiology and Electro-Therapeutics, and established a literature of its own. On these two branches of the science, treatises by the most accomplished physicists and medical men have for years been issued from the Press on both sides of the Atlantic. Week after week new discoveries were made as to the power of electricity to attack and conquer diseases of the most varied and complicated form, so that at the present moment the list it has successfully overcome is far too lengthy to be enumerated here, but will be referred to later on. So long as the employment of the Voltaic Batteries was limited to scientific inquiry and professional use, their clumsy and unwieldy construction formed no obstacle to their employment in the laboratory of the chemist or the consulting-room of the medical practitioner; but when the battery had to be employed in out-door practice, or by the patient himself, the want of a portable generator of electricity, always ready for application and easily put in action, was much felt. This want was soon to be supplied; but previous to my relating how and by whom, let me quote briefly from one of the late Dr. Golding Bird's lectures (page 6) delivered before the Royal College of Physicians in 1849. "Few subjects," he says, "have more frequently or with

greater interest from time to time attracted the notice of the physician, than the nature and application of electricity and its modifications to medicine and physiology . . . I therefore purpose, as the subject of these lectures, to draw the attention of the College to the part played by electricity in a physiological as well as a therapeutical point of view, and hope to show that the functions this agent fulfils in health, and its applications in disease are of far greater importance than have hitherto been considered. Conscientiously convinced that the agent in question is not a less energetic and valuable remedy in the treatment of disease, I feel most anxious to press its employment on the practical physician, and to urge him to have recourse to it as a rational but fallible remedy and not to regard it as one capable of effecting impossibilities. I again say that I shall advance nothing but what has been repeatedly tested under my own observation, purposing to lay before you the result of many years' clinical experience in the wards of Guy's Hospital, and hope to make out a strong case in favour of this too much neglected remedy."

8. While Dr. Golding Bird was thus advocating electricity as a curative of disease, I was successfully engaged with various electrical inventions—namely, electro-motive power, telegraphs, electro-quantity batteries and magneto-electric machines* (see pages 68, 69, and 70), &c., &c. I, however, undertook, at the instance of some medical friends in Vienna, to improve the apparatus employed in medical electricity; and soon found that a radical transformation of the Voltaic pile, allowing of the administration of galvanic currents

* **Patents, Nos. 12,899 and 13,933 (*Old Law*).**

in a *permanent* manner by means of a portable apparatus, would alone offer a chance of success. This new Electro-Therapeutic method, favouring circulation, nutrition, and secretion, very soon proved to be efficacious, even in cases in which the ordinary electrical treatment had failed; and, with but few exceptions, the applications became easy to the patient and convenient to practitioners, who, being relieved from operative treatment, had only to prescribe. The late Dr. Golding Bird, for many years the able Professor at Guy's Hospital, was so impressed with these electrical improvements that he spoke of them, as reported in the *Lancet*, Vol. II., 17, 1851, as follows, the report occupying three columns of the journal:

"The ingenious Galvanic Chain of Mr. Pulvermacher "has attracted so much attention that an account of its "value may prove interesting. . . . Shocks are not "required to develope physiological phenomena or the-"rapeutical effects, as the laborious researches of Dr. "Marshall Hall have long since proved, and it is only "to the *mild continuous voltaic* current that we must "look for a vast development of therapeutical influence."

And in an autograph testimonial he says: "We have in "this ingenious invention what has long been a deside-"ratum—viz., an apparatus of the smallest possible bulk, "capable of evolving a *continuous* or *uninterrupted* cur-"rent of Electricity of moderate tension, and always in "one direction, without the expense, bulk, and great "inconvenience of the Cruikshank trough, or the cell "arrangements. I can hardly recommend Mr. Pulver-"macher's invention too strongly to my medical breth-"ren."

ELECTRICITY, POSITIVE AND NEGATIVE, DEMONSTRATED BY EXPERIMENT.

9. Having proceeded thus far with the subject, it is

now necessary that I should retrace my steps, in order to point out and explain certain phenomena, without a due understanding of which the statements made would be meaningless, and consequently valueless. We will first consider the *attracting* and *repelling* properties of electricity.

These can be illustrated by the following simple experiment: Suspend a light downy feather, or a pith ball, in the air, by means of a thread of silk; then take a glass tube, about three feet long, and one inch in diameter, perfectly dry and clean; rub it briskly with a warm silk handkerchief, and it will soon become electrically excited; on holding it within a few inches of the feather, the feather will fly up and remain adhering to it. On withdrawing the tube, and again bringing it near to the feather, it will no longer be attracted, but repelled. When the feather is attracted by the glass tube, it takes from the latter a portion of electricity, and then suffers repulsion. In a short time the feather loses its electricity in the surrounding air, and is then in a condition to be again attracted by the tube.

We see, therefore, that the feather, having derived a share of electricity from the glass, is repelled by it, and as this repulsion can be proved to be mutual, we learn from this experiment the electrical law, that bodies *similarly electrified*—or in other words, charged with electricity of the same kind—*repel* each other, whilst bodies dissimilarly electrified *attract* each other. If this experiment be repeated with a stick of sealing-wax, chiefly composed of resin, and rubbed with dry flannel, instead of the glass tube rubbed with silk, the same results will be produced. But on presenting to the feather, first the excited glass and then the excited sealing-wax, we shall observe this remarkable fact—while the *glass* will *repel* the feather, the *sealing-wax* will *attract* it, and *vice versâ;* or, if the glass be held on one side of the feather, and the sealing-wax on the other, the feather will be alternately attracted and repelled, and swing to and fro between the two electrics. By this latter experiment it will be seen, that there must evidently be some difference between the electricity of *glass* and the electricity of sealing-wax, or *resin*. That from glass is termed VITREOUS, and from resin, RESINOUS; they are also known as POSITIVE and NEGATIVE; a body electrified by vitreous electricity is said to be in a *positive* state, and that by resin, in a *negative*.

Again, if we bring near to the feather that part of the handkerchief which was used to rub the glass tube, after it was repelled by the latter, the feather will be attracted.

These phenomena are identical with those observed in connection with the electricity drawn from the glass disc, and from the cushions of the frictional electrifying machine, now disused in Electro-Therapeutics.

I now come to speak of—

10. INSULATORS AND CONDUCTORS.

All solid substances admit of being electrically excited by friction; but in a large number of them the electricity disappears as fast as it is generated. If a rod of metal be held in the hand, and rubbed with silk or flannel, it will exert no action on the suspended feather, because all the electricity, as fast as it is formed, will flow or be CONDUCTED through the body into the earth. But if the rod of metal be provided with a handle of glass, or resinous matter, and then rubbed with a piece of dry, warm flannel, it will attract or repel the feather. It is, therefore, evident that the handle of glass or resin INSULATES, that is, separates the metal rod from surrounding objects, and prevents the electricity from quitting it. Metals are by far the best CONDUCTORS; and next to these in conducting power are well burnt charcoal, plumbago, acids (concentrated and diluted), saline fluids, water and moist vegetable substances, living animal matter, flame, smoke, soot, and steam. But these vary greatly in their power of conduction. Among the more important *conductors* are earth and stone, dry bones, chalk and lime, marble, and damp paper; less perfect conductors are dried vegetable and animal substances, parchment, leather, feathers, oil, and fatty substances, pitch, and silk; the worst *conductors* being fur and hair, dry air and gases, steam at a high temperature, glass, gems, amber, resins, and brimstone. By taking these substances in the reverse order we have a list of ELECTRICS, or INSULATORS; those which are used for exciting electricity being the worst CONDUCTORS, and therefore the best INSULATORS, and *vice versâ.* But we must bear in mind that no substance is absolutely a *perfect* insulator or a *perfect* conductor; the insulating and conducting powers differ only in degree, but the extreme differences are so great, that the substances first mentioned may be regarded as conductors and non-electrics, and those last mentioned as insulators and electrics. The difference between a good conductor and a bad one is illustrated by the fact that iron wire conducts electricity four hundred million times better than pure water; or, in other words, the electricity meets with no more RESISTANCE in passing through an iron rod four hundred millions of inches in length, than in passing through a column of water of the same diameter only *one* inch in length. Thought is the quickest of all travellers, and electricity stands next in regard to speed. It has been proved by experiment that electricity moves through a copper wire, one-thirteenth of an inch in diameter and about half a mile in length, at the rate of five hundred and seventy-six thousand miles in a second of time, a velocity greater than that of light.

It must be borne in mind, that the electricity produced by the excitation of glass or resin has been called, from its source, *Frictional Electricity,* which, though small in quantity, is of high tension. In fact, its character, in comparison with that developed by the Voltaic pile or battery, may, as an illustration, be represented by steam at very high pressure compared with that at low pressure. I will now draw attention to Galvanism or Voltaism, by which electricity is developed without the action of friction.

11. If two polished metallic discs—one of copper and the other of zinc, about three inches or so in diameter, and each provided with an insulating handle—are brought into contact (holding them by their handles), and are then separated, and brought successively into contact with the collecting plate of a condensing electroscope, attraction or repulsion of the gold leaves is at once visible, however small the electric power may be. The *zinc* plate will be found to be slightly charged with *positive*, and the *copper* plate with *negative*, electricity.

12. This instrument (the electroscope) consists of

two slips of gold leaf (instead of the feathers mentioned before in the fundamental experiments) suspended within a glass jar from a wire passing through a glass tube, by which their complete insulation is effected.

13. The electric effects thus obtained were considered by Volta to be due to a peculiar electro-motive force, under which metals by simple *contact* have a tendency to assume opposite electrical states; and this view has been supported in recent times by a brilliant array of profound electricians.

14. On the other hand, a powerful mass of evidence against it, and in favour of the theory that the source of power is *chemical action alone*, has been brought by numerous *savants*, including Wollaston, Becquerel, Schoenbein, Faraday, Grove, and De la Rive. Assuming that electricity is developed by the contact of dissimilar metals, it is easy to understand that the increase of chemical action must give rise to an increase of electrical force.

15. If, for instance, we take a plate of zinc and copper, or platinum and zinc, and whilst immersing them in water, cause the metals to touch each other, a Galvanic circle will be formed, and the water will be slowly decomposed into its elementary gases—*oxygen* and *hydrogen*. The oxygen will chemically combine with the zinc, and at the same time a current of electricity will be transmitted through the liquid to the platinum, on the surface of which the liberated hydrogen will make its appearance in the form of minute gas bubbles. The electrical current passes back again into the zinc at the point of its metallic contact with the platinum, thus keeping up a continual current; hence it is called *a Galvanic circle.* The moment the circle is broken the

current ceases, but is immediately renewed by again closing it and re-establishing the metallic contact.

SINGLE LIQUID BATTERIES, BY WOLLASTON, SMEE, ETC.

16. If the water be mixed with a little acid (say sulphuric, hydrochloric, or acetic) these phenomena will occur in a more intensified degree. An increased quantity of the gas bubbles will become visible on the platinum (the negative metal), and a greater quantity of electricity will also be produced. But, from the fact of a portion of the hydrogen adhering tenaciously to this negative plate, the current is still unduly weakened, in consequence of the hydrogen giving to the *negative* platinum the polarity of a *positive* metal (zinc). This effect is termed "polarisation" of the negative element.* Voltaic arrangements of this kind, charged with a single liquid, are termed, after the names of their inventors, Wollaston, Smee, Walker or Stoerer Batteries. For the reason above indicated, they are inconstant in their action, and are therefore advantageously replaced for some purposes by arrangements yielding an electric current of greater constancy, obtained by the oxidation of the hydrogen as it is liberated from the surface of the negative element.

17. In the combinations of Daniell, Grove, Bunsen, Becquerel, &c., the constant current is obtained at the cost of increased complication, and the use of highly corrosive (oxidising) fluids, and they are consequently quite inapplicable for ordinary use, and especially for medical purposes.

18. We now return to the subject of MEDICAL ELECTRI-

* In Pulvermacher's Chain-Bands, &c., this "polarisation" is prevented by the free access of atmospheric oxygen to every portion the negative surface.

CITY, presuming, after the explanations given, it will be more easy to comprehend what we have now to describe. Scores of machines have been invented for the medical application of electricity, but, for general practical use, none have secured so much esteem or been employed so extensively as Pulvermacher's Patent Voltaic Chains and Bands, which, after many years of hard study and labour, the inventor has at length brought to a high state of perfection. The characteristic which distinguishes them from the batteries formerly produced is, that they are pharmaceutical in their nature and application—that is to say, are as portable and easily used as any ordinary external remedy; whilst the old appliances were essentially physical instruments, suitable for a philosophical cabinet, but unfit for general and popular use. Everyone who has at heart the welfare of mankind, and anyone who has suffered or is suffering from ailments, of whatever nature, and is desirous of relief or cure, will be interested in the following details relating to these Chains and Bands, more especially as they are given in a simple and popular form, without confusing the reader with any abstruse scientific investigation.

10. PULVERMACHER'S VOLTAIC CHAIN BATTERY

Is composed of 120 links, or, properly speaking, *elements*, mounted in two separate chains, and is bright and brilliant in appearance, and as neatly and carefully finished as a piece of jewelry. When not charged, it is as inanimate as an ordinary chain; but let it be passed through a diluted acid—vinegar or acetic acid—and it immediately becomes an active and powerful agent. When the dissimilar ends or "poles" of these two pieces of excited chain are grasped in the hands, and their free ends allowed to touch each other, on the instant of contact a contraction of the muscles takes place, and a sharp shock is experienced.

CONTINUOUS AND INTERRUPTED CURRENTS.

DESCRIPTION OF INTERRUPTOR BY PULVERMACHER.

20. When the current derived from a voltaic arrangement or battery is allowed to circulate without interruption it is called "continuous current;" but when its continuity is broken by an interrupting medium, it is called "intermittent current." This latter is most conveniently effected by means of PULVERMACHER'S INTERRUPTOR, which consists of a little glass tube containing a vibrating spiral of copper wire; this is attached to the free ends of the Chains and agitated, when a succession of vibratory shocks is distinctly felt through both arms. These shocks can be increased in strength and intensity to an unbearable degree, by connecting several Chains by their dissimilar poles; but by lessening the number of links or elements, the strength and intensity can be so reduced as to be easily bearable by a child.

If the INTERRUPTOR is detached, and the swivels at the free ends of the Chains hooked together, no shocks are experienced, no matter how many links or elements are introduced into the circle. It might be supposed, by the absence of any shock after the removal of the INTERRUPTOR, that Electricity ceased to be developed, and that the Chain had exhausted its power. Such is not the case, as is evinced by the insertion, in lieu of the INTERRUPTOR, of a little glass test tube.

21. The VOLTAMETER,* or measurer of Voltaic Elec-

* A glass tube containing water, into which pass two pieces of platinum wire, which are respectively connected with the poles of the Chain.

tricity, is connected with the Chains in the same manner as the INTERRUPTOR. As soon as the connection is made, the Electricity travels through the water from one platinum wire to the other therein contained, and in so doing decomposes the water—that is, causes it to separate into its chemical elements—oxygen and hydrogen. By closely observing these platinum wires in the Water-Decomposer, gases will be seen rising in minute bubbles to the surface of the water; but it must be borne in mind that the body forms part of the electric circle. This experiment shows in a manner which cannot be doubted, that the Electricity is *actively traversing the circle,* although *it does not make its presence felt by any shock.*

Another way of testing, which, however, does not admit of ocular demonstration, but none the less proves that the continuous Electricity of the battery passes through the human body, *though no shocks be felt,* is its stinging effect on the sensitive nerves. This is occasioned by the development of heat, scientifically named calorific action, which always accompanies to a lesser or greater extent such electric currents, especially on the surfaces of the body where the nerves are more or less prevalent, and therefore proportionately sensitive; as, for instance, the palm is not so susceptible as the back of the hand. When alluding in a previous portion of this treatise to the weaker or stronger shocks obtainable by few or more links, or elements, inserted in the circle, it was shown that the *differences between these effects* must be the result of *differences in the causes* which produce them. And here for a moment we return to the subject of CONDUCTORS.

22. We have already stated that a vast difference exists in the conductibility of certain substances, and that living animal matter is, as compared with metals, an inferior CONDUCTOR; thus, in the case of the human frame, it is found that the different parts of which it is composed vary greatly in this respect. The hair, nails, bones, and nerves are, as conductors, greatly inferior to the muscles or flesh; and these in their turn are inferior to the blood and other liquids contained in the body. Conductibility must, however, be regarded in relation to the TENSION or density of the electricity used. By TENSION is meant the power of a current to force its way through any medium of *inferior* conducting capacities.

23. A body inserted in a circle, which would resist the passage of a current of *low tension*, would offer but a slight *resistance* to the passage of one of *high tension*. Electricity produced by friction is generally endowed with high tension, whilst that obtained from the Voltaic battery is of low tension; or, as already stated, frictional is to Voltaic electricity as high-pressure to low-pressure steam. A compound battery may vary in its tension according to the number of elements voltaically combined together. For instance, whilst one single element is of *low* tension, a voltaic combination of such elements will increase the tension proportionally to the number of additional elements. If we want to send a Voltaic current through those parts of the human frame which offer great resistance—say from one hand to the other, or from the hands to the feet, which present a small section of greater length in comparison with the rest of the frame—a battery of a large number of elements is required. If the current is to traverse a short distance through

a part of the body, say across the trunk, a battery with fewer elements will produce nearly as much effect as a battery of many more elements. From these remarks one may readily perceive that a Chain of fifty elements will produce feebler shocks than one of a hundred, the power of these shocks depending on the part of the body through which they have to be sent. Let it also be borne in mind that the intensity of an electric current may be increased by the enlargement of the metallic surfaces of which a Voltaic element is composed, and also by increasing the acidity of the exciting liquid by which it is set in action. That such will be the result can be plainly shown by the increase of the gaseous column in the Voltameter, when it is placed in connection with the enlarged and more highly excited elements.

24. What has been stated with reference to an *imperfect* conductor, such as the human body, has an equal application to the *best* conductors—namely, metals. Take, for example, a wire: the thinner and longer it is, the greater will be its resistance to the electric current, and the higher will be the electrical tension required.

Here we must briefly refer to a wonderful application of Electricity, which hardly comes within the pale of this Treatise.

It was known from an early period in the history of Electricity that a certain connection exists between it and Magnetism. Many noted electricians had searched into the matter, but it is to Professor Œrsted, of Copenhagen, that the world is indebted for the discovery of the great and useful fact (established in 1819)—that an electric current travelling along a conductor placed parallel to a suspended magnetic needle, deflects the latter, so as to place it in a transversal direction, or across this conductor; thus proving that *electricity in motion* will impart *motion* to the magnetic needle. He also found, that the nature of the conducting medium is immaterial to the result; whether the Voltaic circle is completed through a metal, or a conducting liquid contained in a glass tube, the magnetic needle is equally moved. Deflection takes place in one direction when the needle is placed *over* the conductor, and in the opposite direction when placed *under* it.

25. Again, winding helically a metallic wire, covered with an insulating envelope, over a hollow (flat) bobbin, an electric current passing through this wire will exercise on a magnetic needle suspended inside the bobbin a degree of deflection *proportionate to the strength of the current* and the *number of rounds* or turns of wire. On this electro-magnetic principle the GALVANOMETER and the TELEGRAPH are constructed. Most undoubtedly, through this discovery, Professor Œrsted may be truly considered the father of telegraphy.

ELECTRO-MAGNETISM, INDUCED ELECTRIC CURRENTS, TEMPORARY AND PERMANENT MAGNETS.

26. If a rod of soft iron be placed within the hollow bobbin, it becomes itself a magnet during the time that the current of electricity circulates in the wire coiled round it, but the moment the current ceases, the rod loses its magnetism; hence the expression *temporary magnetism*, in contradistinction to the *permanent magnetism* which a rod of hardened steel, after being magnetised, represents. The illustrious Faraday, however, completed the great discovery of Professor Œrsted by patient researches, and revealed another great fact in

electro-magnetism—namely, that, when one of the poles of a magnet is moved past (across) a wire, the ends of which join in a conducting circle, a momentary or *intermittent* current of electricity is *induced* in the wire. Hence the term "induction current." This fact may be demonstrated by inserting in the circle the galvanometer just described. Deflection of the magnetic needle in the galvanometer occurs as often as the pole of the magnet is moved across the conducting wire; hence we may say that *magnetism in motion produces electricity, and electricity in motion produces magnetism*, as shown above. On this great physical principle are constructed electro-magnetic and magneto-electric machines, the use of which has benefited medical electricity ever since their discovery. The current of electricity thus induced is composed of a series of intermittent (momentary) impulses, produced by a succession of inductions at short intervals; and the strength of the current depends on the number of windings in the inductional coil, and the intensity of inductive power used. The multiplication of these windings causes an increased tension analogous to that produced by an increased number of elements in the Voltaic pile.

27. In sending a message to America by a submarine wire, the electric current meets with great resistance to its passage along the cable wire; yet, notwithstanding this opposition, a chain of six *small elements* will furnish a current of such power as to be able to act on the magnetic needle in the instrument on the further side of the Atlantic, and by this means deliver the message in its own conventional style. Again, if we want to produce the greatest magnetic attraction by means of electro-magnets, formed of coils

of *thick* insulated copper wire surrounding the core of soft iron, *one single element* will suffice for magnetizing the iron core to its maximum power; but the magnetic power thus produced depends on the proportionate electric quantity derived from the larger or smaller metallic surfaces of the elements, and the stronger or weaker acidity of the exciting liquid. When the electric current of a single Galvanic element is applied to the human frame, however, or brought in connection with the Voltameter,* its effect will be feeble, in consequence of all the electricity produced by a single element not having the power (tension) to overcome the resistance which it meets in these bad conductors.

PULVERMACHER'S MEDICO-GALVANIC SYSTEM.

28. The reader will now be able himself to answer the question often put forward by the uninitiated in electricity, "Which kind of battery is the best for medical use?" He will also have gathered, no doubt, from the facts stated and the experiments described, that the battery with the largest elements, when these elements are but few, is feeble in its action on such parts of the human body which offer an overwhelming resistance to the passage of electricity; whereas a battery with small-sized but numerous elements, in such application, will produce physiological results both striking and startling.

29. It was these facts, which prompted the study and ardent research of years, and gave birth to the Pulvermacher System of Medical Batteries in the shape of Chains and Bands, which evidently inaugurated a new

* See page 23.

era in the history of Therapeutical Electricity. The description of Pulvermacher's Flexible Medical Batteries is to be found in the following leading modern works on Natural Philosophy and Electricity:—

30. Electricity and Magnetism. By Prof. Becquerel ... 1853
Guy's Hospital Reports, p. 107 ... ...
Elements of Therapeutic Physics. Dr. Heidenreich, p. 248 ... 1854
Dr. T. Pereira, F.R.S., &c. Materia Medica, p. 53, ed. IV. .. 1854
Local Electrization. Duchenne, p. 39 1855
Elements of Physics. Prof. Pouillet, p. 694 1856
Lancet, Vol. II., xvii., 1851, and No. I., Vol. II. 1856
Exposé de l'Application de l'Electricité, 2nd ed., Vol I., pp. 115 to 239. Du Moncel 1856
Application of Electricity. By Viscount Du Moncel, pp. 237 and 239, II. 1853
Becquerel on Electricity, its Application in Medical Treatment, p. 36 1857
Bulletin de l'Académie de Médecine, Paris. Vote of thanks to Inventor. Vol. XVI., p. 13 1857
Electricity and Medical Treatment. Dr. O. Kowalewski, Vol. I., p. 121 1857
Treatise on Electricity in Theory and Practice. De la Rive, Vol. III., p. 604 to 609, ed. I., ... 1858
Phénomène de la Nature. Valerius, p. 239, Vol. I. 1858
Manuel d'Electrothérapie, p. 84, ed. I. Tripier 1861
Electro-Physiology and Electro-Therapeutics. D. Garratt, of Boston, p. 117 1861
Epileptic and Convulsive Affections. C. B. Radcliffe, M.D., p. 180, ed. III.... 1861
Medical Use of Electricity. Garratt, p. 117, ed. II 1861
L'Electricité appliqué au Traitement des Maladies. Par Dr. Desparquets 1862
De l'Electricité de l'Action des Eaux Minérales. Scoutetten ... 1864
Dictionnaire de Médecine, de Chirurgie, &c., ed. II., p. 743. Littré et Robin 1865
Guide Pratique du Doreur. Roselenr, p. 88, ed. II. 1866
Chronic Diseases. Dr. John King, Professor of Obstetrics in Cincinnati 1867
Traité Elémentaire de Physique, 3rd. ed. P. A. Daguin, pp. 408 and 409 1867
Compendium, p. 231 and 232, ed. III. 1868
Electricity, Magnetism, and Acoustics. Lardner and Foster, p. 304 1868

L'Electricité. J. Baille, pp. 201 and 202 1868
National Philosophy. Dr. Lardner, p. 304 1868
Pathology and Therapeutics. By Mr. Wunderlich, Vol. I., p. 113 1868
Gazette des Hopitaux, Paris, 27th July 1852
Althaus, p. 302 and 303, ed. II. 1870
L'Electricité appliqué à la Thérapeutique Chiruigicale 1870
Clinical Electro-Therapeutics. Dr. Allan Hamilton, M.D. ... 1873
Medical Surgical Electricity. Drs. Beard and Rockwell, p. 137 ed. I. 1871
Traité Elémentaire de Physique Médicale. Wundt, p. 571 ... 1871
Popular Natural Philosophy. Ganot, p. 831, ed. 7 1872
L'Electrothérapie dans les Maladies Genital et Urinaire, par le Dr. Deloume 1872
Practice of Medical Electricity. Powell, p. 20 1872
Académie de Science, Paris, extract reported in "The Cosmos"
Medical Electricity. Tibbits, p. 20 1873

PRINCIPLES ON WHICH PULVERMACHER'S CHAINS ARE CONSTRUCTED.

31. The fundamental principles upon which these Chains and Bands are constructed are as follows: *First*:—To place and apply the galvanic metals in such a way, that when they are set in action by a supply of the exciting liquid, the latter, instead of surrounding and *submerging* it as in a bath, should be *absorbed*, and a small portion retained between the zinc and copper surfaces, which are placed in close juxtaposition to each other. By this arrangement the cumbrous trough, cells, and jars used in the ordinary batteries as receptacles for the acids, in which the galvanic metals are placed for excitation, are rendered superfluous. *Secondly*:—To combine the galvanic elements into a compound Voltaic battery, by attaching them in such a manner to each other that they may be *movable*, without in the slightest degree interfering with the permanent contact between the copper of one element and the

zinc of the next. *Thirdly*:—To render instantaneous the charging of a great number of elements thus flexibly joined together, by simply passing the combination or compound Chain-Battery through the exciting liquid. *Fourthly*, to allow *access to the atmospheric oxygen* simultaneously with the exciting liquid, so as to counteract to an extent the *polarization* of the *negative* metal, and thus give to the current *constancy of action* for a given time, notwithstanding the small quantity of exciting liquid absorbed by capillary attraction.

To carry out the foregoing principles, the plan of combining the elements as links of a Chain suggested itself; and to its practical realisation are due many advantages, such as portability, promptness in setting the Chain-Battery in and out of action, a simple and easy *graduation* and *regulation* of the current, both in its *continuous* and *interrupted* state, and, *above all*, comparative cheapness. The progress in Medical Galvanism, as shown by this invention, was so thoroughly understood by those competent to judge in such matters, that it is not to be wondered at that these Chains, from the date of their first appearance, received the approval and admiration of the scientific world.

32. To refer only to the Académie de Médecine of Paris, the report on these Chains, read at the meeting on April 1st, 1854, contains the following valuable testimony:—

"The Voltaic Chains of Mr. Pulvermacher are really "a most wonderful apparatus. . . . They are more "portable and cheaper, which are two indispensable con-"ditions in an apparatus of this description, in order to "make the application of Electricity more general, and "to a certain degree popular, which is certainly very de-"sirable in the interest of patients, as well as that of the

" profession. In this respect the Chains of Mr. Pulver-
" macher will have a great future . . . The com-
" mittee beg to propose to the Academy to address their
" thanks to Mr. Pulvermacher for his most interesting
" communication. Adopted."—*Bulletin de l'Académie*, Tome XVI., No. 13.

Were it not for limited space, many similar quotations could be given, rendering self-praise superfluous. I cannot, however, refrain from giving a few more extracts, considering it preferable to let others speak, who, by their position and judgment, are competent to appreciate the merit of these inventions. I will only cite the opinions of men who have risen to the pinnacle of their profession, such as Sir CHARLES LOCOCK, Sir HENRY HOLLAND, Sir WILLIAM FERGUSSON, Sir RANALD MARTIN, &c., &c., quoting the words of their written Testimonial, dated the 9th of March, 1866 :—

" We have much pleasure in testifying that Mr. J.
" L. Pulvermacher's recent improvements in his Voltaic
" Batteries and Galvanic Appliances for medical pur-
" poses are of great importance to scientific medicine,
" and that he is entitled to the consideration and
" support of every one disposed to further the advance-
" ment of real and useful progress."

Dr. KING, Clinical Professor of Obstetrics in Cincinnati, eminent amongst the medical practitioners of America, says, in his standard volume on Chronic Diseases, page 76 :—

" The instruments for the purpose of making thera-
' peutical employment of the Continuous Galvanic Cur-
" rent are Daniell's Batteries and Pulvermacher's Im-
" proved Chains. In those cases where it is desirable

"to produce a Continuous Current of Galvanism, and "without the intervention of conductors or electrodes, "there is no instrument superior to *Pulvermacher's Im-* "*proved Galvanic Chains.*" [Here follows a description of the construction, after which he continues further.] "When these Chains are immersed in vinegar, they "are impregnated with the fluid, and the action of "the Battery is excited, and a steady current of Gal- "vanic Electricity is kept up."

33. The inventor has always considered it a duty incumbent upon him, in spite of all the opposition and drawbacks to which every new invention is subjected, to exert himself in making his Chains as perfect as possible; in order to satisfy the wants of the public on the one hand, and on the other to disarm the interested criticism of those who, to conceal their partiality, are always ready to pick out minor imperfections. It would be idle, though perhaps amusing, to enumerate the absurdities and inconsistencies of the denunciations put forth, and the objections advanced; here, however, is a specimen, dedicated to all who like fair play and love truth. Medico-electric specialists speak of the "*inconstancy* of the electric current," forgetting that the facility in recharging these Batteries compensates amply for the relative inconstancy complained of. It must also be remembered that an application of galvanism rarely exceeds twenty minutes, during which time the current of these Chains continues its uniform action, in spite of the small quantity of exciting liquid absorbed. One might as well complain of the inconstancy of heat produced by a spirit-lamp compared with that generated by a furnace, and unadvisedly prefer to make use of the

latter for such purposes where a spirit-lamp would be more practical. As to the portable Chain-Band, for diffusing a *gentle* galvanic current, its *raison d'être* and practicability must recommend it to patients.

Some amongst the medical profession say, "It is too simple, therefore too trivial for our use." Others, "It is too complicated for our daily practice." Some patients, having overcharged the Chain, find its application irksome, in consequence of the stinging sensation produced; whilst others charge it too feebly, or not at all, and complain that they feel no effect. Another, again, takes umbrage at the words "Electricity is Life," adopted years ago as a distinctive trade mark or motto, and brings an accusation of exaggeration; as though it were intended to convey the idea that Electricity is life itself, whereas the phrase was used merely as a figure of speech, similar to that which says "Heat is Life," "Movement is Life," and, as it is expressed with reference to the blood, that it is "the life" of the animal.* Then we have that legion of sceptics, and those uninitiated in electrical science, who avoid even the trouble of investigation, and find it more to their purpose to deny its efficacy than to give the invention a trial. When we consider these drawbacks, which are but a fraction of those with which this invention has had to contend, it is a matter for

* This biblical statement, it may be interesting to observe, is borne out by Electro-Physiology, which shows that the electro-chemical properties of the blood originate those electric currents which, in their turn, by means of the nervous system, govern the production and circulation of this nutritive fluid.

wonder that it has been able to acquire that popularity in England and abroad which thirty years of useful existence has procured for it. The groundless objections before referred to have even prompted speculators to launch *quasi*-electric contrivances, as a substitute for Pulvermacher's Patent Appliances. Nay, more, infinitesimal Magnetism, hidden in all sorts of hosiery, has been put forward as a competitive means of supplying the electro-curative agency, under the plea of superior comfort, but necessarily at the expense of efficacy and the sufferer's purse. It must be remembered that, according to the opinion of the late Professor Faraday, Magnetism is devoid of any physiological action, and it is therefore preposterous to claim for it any curative effects whatever.

34. Being aware of the deficiencies which formerly existed, the inventor has discovered the ways and means of removing them, a fact to which his present improvements and additions bear witness. Yet, to please everybody is no easy task; and to make one and the same remedial appliance answer for a variety of ailments, constitutions, and temperaments is generally considered next to an impossibility. In the case of Electricity as a remedial agent, however, the immortal John Wesley, in his celebrated work entitled "Primitive Physic," observes: "If there be a panacea (*i.e.* a universal remedy), Electricity is certainly the most deserving of that title." But great and extraordinary as are the cases of cure recorded by that remarkable man, and the men of his day, they are as nothing in comparison with what has been accomplished since the healing properties of Voltaic Electricity, and the much more simple

and effectual means of applying it, have been better understood.

The purely scientific origin, and the complex nature of the Medico-electrical appliances formerly in use, made their application alike difficult to patients and physicians who were not versed in the science. But the great success which attended the use even of these unwieldy and complicated Batteries rendered it vastly important, and, indeed, indispensable, that apparatus should be contrived, at once simple in construction, easy of application, and at the same time of such low cost that their vast and valuable properties could be accessible to every patient, either with or without the intervention of a Medical Electrician. When, twenty-seven years ago, Mr. PULVERMACHER invented his first patent Hydro-Electrical Chain, a great step towards popularising this mode of treatment was made, and new fields for improvements opened, which the Inventor has never ceased to cultivate. The great simplicity in the construction of these Chains, and the easy manner in which either a mild or powerful action can be obtained at a minute's notice, has made them alike valuable to the patient and to the medical practitioner. Their great portability has given birth to a new system of Medical Electricity, of most effective application, which Mr. PULVERMACHER may honestly claim as *his own*—namely, the GENERALISED *permanent* application of a portable and flexible voltaic arrangement *direct* to the part affected.

By accomplishing this simple method, of diffusing with readiness the electric current, he has succeeded not only in reducing the means of applying the varied degrees of power of electricity into the smallest possible bulk and lightest weight, but also in securing unlooked-for comfort in self-application, so as to suit the most sensitive patients, delicate constitutions, and tender age.

This system, misunderstood and opposed at first by Medico-electric specialists, was most successful in its curative results, and is now finally countenanced and appreciated by its former opponents.* The extensive use which has been made of these galvanic appliances during the last quarter of a century, has furnished Mr. PULVERMACHER with the necessary experience for the study and realisation of the improvements which he introduced, and thus justified the encouragement he received from the Academy of Medicine of Paris, and the *élite* of the medical profession in England and abroad. Study and observation have shown that the *kind* of Electricity, the *degree* and *duration* of power to be administered, are of as much importance as the *mode* of application, so that treatment be carried out in accordance with the universal laws governing cause and effect.

35. A long and extensive use of these appliances has shown that successful treatment by Electricity depends as much upon the proper quantity—or *dose*—of the electrical current, as does ordinary treatment upon the quantity of medicine administered. The use of the ordinary powerful electrical apparatus, the current of which must necessarily be reduced to a feeble degree, is of much less advantage in the hands of the patient than are special appliances for the application of weaker currents, which at the same time com-

* See recent publication, "*Graduation et Dosage du Courant Continué*," by Dr. DUCHENNE (Paris, 1873). Also "*Application de l'Electricité à la Médecine*," by Dr. TRIPIER (Paris, 1874).

bine *simplicity, safety,* and *cheapness.* For the easy and yet effectual use of a *mild continuous current* by the patient himself, special appliances were necessary, since it is well known that a voltaic current of Electricity is endowed with a chemical and calorific action; and as a given quantity of caloric (heat) is endurable only when diffused over a large surface of the body, but painful if concentrated upon a given spot, so also with Electricity —if used too strongly, the concentration upon a limited surface becomes irritating and irksome. To remedy this inconvenience it has been necessary to reduce the strength of the current so as to suit the sensitiveness of the patient; but the amount of Electricity often required in certain ailments was thus rendered insufficient. In order to reap the advantage of the *whole strength* of the battery at disposal without any counter-irritation, and be enabled to apply *mild* continuous currents permanently, was a problem which remained to be solved. This led Mr. PULVERMACHER to his latest invention, the patent ELECTRODES and SELF-SECURABLE ELECTRODION, which will be found a *great boon to many sufferers* from a class of ailments where a minimum dose of Electricity is to be directed to a part of the body subjected to pressure—as beneath a stay, boot, or orthopædic instrument, &c., &c. In cases, however, where a more powerful current is required, the ELECTRODES serve as a ready means of diffusing the same over *a large surface,* preventing stinging sensation or cutaneous irritation.

These improvements are calculated alike to please the practitioner and the patient. For the use of the former, Mr. PULVERMACHER has now brought out a new or improved flexible Battery, which can be charged instantaneously, and used even while being carried in the pocket, thus furnishing a vade-mecum for out-door practice. Though small in compass, it will give off a *greater amount* of Electricity and generate currents of *greater constancy* than any of the small element batteries hitherto produced. Its power and convenient size renders it invaluable in surgical operations, such as electro-cautery, &c.

Another great improvement is the new *Galvano-Piline Bands of constant action*, provided with an *insulating wrapper* of soft material, and capable of producing large quantities of Electricity, which can be conveyed to the system. It thus affords, in addition, the comfort of a warm and dry application, in connection with the ELECTRODES; whereby intensity currents can be administered to the body *without the necessity of the Bands being worn next the skin*, thus admitting of their being re-charged without the patient undressing. The perfection of comfort being now secured by Mr. PULVERMACHER'S latest improvements, the most delicate patient can henceforth avail himself of the efficacy of this electrical treatment, without any inconvenience whatever, and at little expense.

36. Those readers whose interest and attention has been secured thus far, and who have felt gratified by the various facts with which they have become acquainted, will undoubtedly perceive the importance of the subject when they become aware of the extent of the curative powers with which Electricity is endowed. They will naturally be impelled to enquire, "To what kind of ailments are these Electric Chains and Bands applicable?"—the answer to which is authentically given in the following extracts:—

The Rev. JOHN WESLEY, M.A., in his "Primitive Physic," written in 1755, when the electrifying machine by friction was the only means known, says :—

" One remedy I must aver, from personal knowledge " grounded on a thousand experiments, to be far superior to " all the other medicines I have known—I mean Electricity. " I cannot but entreat those who are well-wishers to mankind " to make full proof of this. Certainly it comes the nearest " to a universal medicine of any yet known in the world."

Page 116.—" Electrifying is proper in many cases—St. " Anthony's fire, blindness, blood extravasated, bronchocele, " burns or scalds, coldness in the feet, contraction of the limbs, " convulsions, cramp, deafness, falling sickness, feet violently " disordered, fetous fistula, lacrymotis fits, flooding, ganglions, " gout, headache, imposthumos, inflammation, involuntary " motion of the eyelids, king's evil, knots in the flesh, lame- " ness, lockjaw and joints, leprosy, menstrual obstruction, " ophthalmia, pain in the stomach, palsy, palpitation of the " heart, wasted limbs, rheumatism, ringworm, sciatica, " shingles, sinews shrunk, spasms, stiff joints, sprains how- " ever old, surfeit, swellings of all sorts, sore-throat, tooth- " ache, ulcers, wens, wastings, and weakness of the legs. " Nor have I known one single instance wherein it has done " harm."

The learned M. DU MONCEL in his *Exposé de l'Application de l'Electricité*, 1857, page 375, says :—

" If we are to refer to the numerous experiences made by " MM. HUMBOLDT, ALDINI, LABAUME, FABRE, PALAPRAT, " RITTER, BICHOFF, MAJOU, ROSSI, GRAPENGIESSER, BAUDE- " LOCQUE, BERMUNDI, PRAVAZ, LE ROY D'ETOILE, ANDRIEUX, " FOZEMBAS, MATEUCCI, BAILLY ET MEYRAUX, PREVOST ET " DUMAS, RECAMIER, TAVIGNOT, &c., *galvanisation* would " have effects analagous to those of electricity, and gene- " rally of more striking efficacy on living beings in its appli- " cation in disease; consequently, in applying it one may " accelerate circulation of the blood, increase perspiration, " promote excretion of certain fluids, and eliminate alvine " matter, disturb the limpidity of the bile and urine, and

" indeed cure or relieve a host of diseases, such as rheumatic " affections, sciatica, gout, asphyxia, certain kinds of mad- " ness, spasms, ruptures, inflammatory tumours, dyspepsia, " liver complaints, disease of the intestines, disease of the " kidneys, diabetes, disease of the bladder, paralysis, para- " plegia, hypochondria, asthma, atrophy, phthisis, hydro- " celes, scrofula, mercurial diseases, varicoceles, sarcoceles, " amenorrhœa, dysmenorrhœa, uterine deviation, tic-doulou- " reux, goitres, swellings, muscular relaxation, ophthalmia, " amaurosis, deafness, intestinal invagination, labour pains, " hæmorrhage, &c.

" According to the before-mentioned physicians, *galvani-* " *sation* would still have a peculiar character which would " make it available in special applications; for instance, " muscular contraction would always be effected when the " direction of the current is reversed—that is to say, by " changing the negative to the positive pole, from which " result one might, by changing the direction of the current, " cause vomiting, or alvine evacuations from the intestinal " tube, or in other terms, determine the peristaltic or anti- " peristaltic movement of the digestive tube. On the other " hand, *galvanisation* would react in a pre-eminent degree on " the retina by producing luminous impressions, would " stimulate the spinal cord, even nervous fibres in a sepa- " rated state, the ganglionic and lymphatic system, would " modify sensations under certain circumstances, and above " all taste and smell, could, moreover, be employed as em- " menagogues. When the battery is powerful the applica- " tion of *galvanism* causes rather strong physiological con- " tractions, without shocks, with a sensation of numbness, " without pain, since violent shocks are not peculiar to con- " tinuous currents. When the battery is weak this sensation " of numbness changes into a tingling, not unpleasant, " accompanied by agreeable warmth. The prolonged action " of the current with metallic plates causes a noticeable " irritation of those parts of the skin where they are applied, " as shown by the little white pustules resulting therefrom."

Dr. John King, M.D., Professor of Obstetrics, &c., &c., in Cincinnati, 1867, page 78, says:—

" This Chain gives out Electricity of high tension and in

" sufficient quantity for almost any medical purpose. Six " links will yield sufficient Electricity to decompose water " into its component gases. . .

" These Chains are very useful in many nervous disorders; " muscular debility; hemiplegia; paralysis of children; central or cerebral paralysis; spinal paralysis; neuralgia; sciatica; stiff joints; œdema of the limbs; hysteria; hysterical " paralysis; aphonia; epilepsy; torpid liver; asthma; amenorrhea; dysmenorrhea; spinal irritation; nervous debility; " constipation; deafness not due to actual disease or structural change; rheumatism; dyspepsia; paralysis of the " bladder; chorea; writer's cramp; hysterical cramps and " contractions; loss of smell; loss of taste; impotency, &c."

COMPTES RENDUS (*TRANSACTIONS*) DE L'ACADEMIE DE SCIENCE, PARIS, ON A PAPER READ FEBRUARY 8TH, 1858, VOL. VIII., BY DR. HIFFELSHEIM, LAUREAT DE L'ACADEMIE.

" The resistance which the human body offers is a great " obstacle to currents of small tension; therefore it is necessary to take a certain number of elements voltaically combined, in order to create sufficient intensity to overcome " this resistance. But this electro-motoric force thus produced results in caloric effects which are not only irksome, " but complicates the treatment. In order, then, to avoid " these effects, the elements must be of small quantity, but " of sufficient tension to overcome the resistance of the " body, and to gain by a prolonged application that which is " lost by the limited quantity which can penetrate at a time. " But how can we realise this particular method, if not by the " aid of portable appliances?—I mean flexible, so as to shape " themselves, as it were, upon the body, by the aid of a pile " which, by its disposition into elements of small surface, " offers comparatively a large surface to the weak exciting " liquid, and does not inconvenience the skin. There is but " one pile, though of different forms, which embodies all " these diverse conditions, and that is Pulvermacher's pile, " arranged in Chains, &c. The small elements of which they " are constructed admit of being multiplied at will, and for " the various degrees of intensity, without rendering the " apparatus more inconvenient. It is with this pile, furnishing a permanent continuous current, and of high tension, "that I have performed these therapeutical applications of "electricity in the Hospital Wards under M. Rayer, who, "although he had been for thirty years using a *Trough* "battery, gladly adopted this form of treatment. It is "worthy of notice that the continuous and permanent "current has a marked action on the muscular system. "Without wishing to examine by what means this physiolo-"gical effect is obtained, it is, nevertheless, certain that it is "not by contractions, as one might possibly think, neither is "it altogether by the contraction which results both from an "interrupted current, but the dynamic electricity acting "directly on the various elementary functions of such multi-"form property constituting nutrition."

"In my applications of the continuous current I have "treated NEURALGIA, often without being able to trace its "origin. The diagnostic was founded on the dominant "symptom, pain.

"Four patients suffering from SCIATICA were discharged "cured in a maximum of fifteen days. One of them, an old "man of 74, had several attacks afterwards, but the pain was "quite of a different kind and of greatly diminished intensity.

"What is most remarkable in this new treatment is the "promptitude with which it acts. The patient or physician, "however, should not be discouraged if a complete cure is "not effected in a couple of days, as the instantaneous "amelioration might at first induce them to hope.

"TIC-DOULOUREUX.—A patient entered the hospital one "morning suffering so intensely that he is unable to move his "head. He had been in the same condition for nearly a "month, and of course unable to attend to his work as a "waggoner. He speaks with difficulty, cannot eat, nor raise "his head, which he holds between his hands all the while. "Besides the continuous pain, he experiences at very short "intervals severe shooting pains, which are almost unbear-"able. A Chain of thirty large elements was applied as a "chin-band at ten o'clock in the morning. Two hours after "the pain had so far decreased that the patient moved his "head. At five o'clock, at our afternoon round, we found "him greatly relieved. The next morning, he informed us "that he had slept a little. A second Chain was then

" applied from the forehead over the head to the nape. During the day he was able to walk about, and could masticate his food. The improvement continued, and on the tenth day he only felt occasional twitches, and after a fortnight he had only a local reminiscence of his pains. In three weeks he was perfectly cured, without the slightest trace of suffering.

"A very violent TIC-DOULOUREUX, speedily relieved and almost cured. Another TIC-DOULOUREUX, very promptly relieved, and cured in less than three weeks. A third TIC OF THE FACE was also cured. A NEURALGIA of the FIFTH PAIR, very well described by the patient, disappeared in less than a fortnight. A young girl suffering from HYSTERICAL AFFECTION, with a violent NEURALGIA of the FIFTH PAIR was also cured.

" SATURNINE OF LEAD COLICS almost instantaneously lost their extreme violence in five cases. The colic was completely removed, but the other symptoms, the difficulty of moving and breathing, which in my opinion constitute the second morbid element of this affection, were more persistent. A SEVERE PAIN, ARISING FROM THE FRACTURE OF THE ANKLE-BONE, disappeared almost immediately.

" I have successfully employed this treatment in two cases of RHEUMATISM not accompanied by fever. Two persons suffering from ST. VITUS'S DANCE were at once relieved, and one of them has felt nothing of it since. A complete SATURNINE PARAPLEGIA was greatly relieved on the instant. The treatment was continued by applying the continuous current on one arm, and every day interupting it on the other arm. Almost completely cured in both arms, the patient, after six months' treatment, left the hospital. Another patient, suffering from a similar PARAPLEGIA, was subjected to the two currents with an almost equal amelioration. A third patient suffering from PARAPLEGIA in the upper limbs, which, as in the preceding cases, was of saturnine origin, being subjected to the two modes of electrisation, gained strength in both arms to nearly an equal extent. The interrupted current was obtained by a Chain-Battery and Mr. Pulvermacher's Interruptor. A PARAPLEGIA successively general and complete, limited and fixed, ultimately, in the limbs of a lad about sixteen was cured by this treatment. In the very first days the progress of the

" disease was arrested, though up to the commencement of " the treatment it had been continually getting worse. A " VIOLENT MERCURIAL TREMBLING was promptly relieved. " A very serious PARAPLEGIA of long standing was accompanied " with all the mobility of a juggler. It was relieved in the " action of the will on the muscle. Two cases of GENERAL " PALSY, one in a male, the other in a female patient, were " treated successfully, the former recovered the complete use " of his limbs, his memory, speech, and sleep; the latter re- " covered her sleep, memory and strength.

" *An extraordinary weakness*, owing to a RHEUMATIC " AFFECTION of the SPINAL MARROW, was greatly relieved in " a few days. A case of a similar kind, but not rheumatic " perhaps, was also very considerably ameliorated. All these " cases were submitted exclusively to the continuous current " of Pulvermacher's Chains."

37. If we take into consideration the vast range of the healing powers with which these Chains, diminutive in appearance, but wonderful in the strength of their electrical activity, are endowed, the question will naturally arise in the minds of those who are unbiased, and who hail with delight every step in the march of progress: "What is the *modus operandi* of Electricity on the human frame which produces such a variety of curative effects?" Previous to answering this question, reference must be made to the various general properties, physical, chemical and physiological, possessed by Electricity, which embrace the entire functions of animal life—viz., movement and sensation; that is to say, MOVEMENT, *voluntary* and *involuntary*, of the organs, and *movement* of nerve-fluid, which produce the sensations of pleasure and pain, experienced concomitantly with the impression received by the individual.

38. Fully to enter here into the physiology of life, or rather the vital processes, would be impracticable, as well

as out of place, as the subject scarcely comes within the range of a popular treatise. Notwithstanding, it is incumbent upon me to allude to certain broad facts, which bear on the question we are at present considering, and which, by experimental evidence, show in a most striking manner, that in the animal economy Electricity is the prime motive power for promoting *circulation, secretion* and *excretion,* accompanied by chemical changes of matter during the processes of *digestion, nutrition* and *assimilation—i.e.,* conversion of food into nutriment, and the distribution of the latter for maintaining the balance of waste and repair. Experiments have established the fact that, by the intervention of Electricity, chemical combinations can be formed which, without such intervention, are quite impossible, and take effect only through the aid or by the means of the processes of life. For instance, the gases, nitrogen and hydrogen, never combine by their own chemical attraction or affinity so as to form ammonia; but the latter is abundantly produced by animal life, and also by electrical action. Again, albumen is never converted into fibrine unless by the functions of life, or through the agency of Electricity. Many such instances might be enumerated which would illustrate this special property of Electricity. Another remarkable feature of the influence of this agency is its effect on the *circulation of the blood,* which is demonstrated both to the eye and the understanding, when we place under a microscope the transparent portion of the limb of a frog, which an electric current is made to traverse. If the electric current be *intermittent,* the circulation of the blood-globules, which is distinctly defined by the microscope, will first *slacken* its speed, and then cease, and may even be reversed; but if the current be made to pass in a *continuous* flow, the normal speed of the movement of the blood-globules,—in other words, the circulation,—will be *accelerated.* Here we have two most interesting and important facts illustrated by a single experiment, which—firstly, puts us in possession of the *means of influencing circulation in any part of the animal frame;* and secondly, shows that the influence, so acquired, can be exercised so as to produce either an *increased or decreased movement of the blood.*

39. By drugs taken internally we are capable of obtaining analogous effects, but these effects extend to and take possession of the whole system, and cannot be arrested at any given moment, whereas those caused by the local application of electricity, although exercising an *internal* action, admit in their *external* application of suspension at a moment's notice.

40. There is yet another experiment which most strikingly illustrates the motive power of Electricity, exercised in Endosmosis and Exosmosis. Let a liquid, say water, be placed in a vessel divided into two compartments by some permeable membranous substance, as a bladder-skin, or any other porous diaphragm. It will be seen, that when an electric current is made to pass by means of electrodes through the water contained in both compartments, the level of the water in the compartment into which the positive electricity enters, will be lowered, and the level in the other compartment will be raised; and if the direction of the current be reversed, the fall and rise of the water in the compartments will also be reversed.

41. These facts most positively prove that *electricity acts on the molecules of liquids like a propelling or transporting power*, so as to convey them through porous substances. This action takes place regardless of the static pressure of the water, and quite independent of those movements of matter which occur in the vital economy.

There are various other phenomena connected with the operative powers of Electricity, but to recount them all would occupy too much time and attention. However, I cannot refrain from referring to a most vital point in relation to the *causes* and *effects* attending the application of Electricity to physiological and curative purposes.

THE HEALING PROPERTIES VARY WITH THE VARIED APPLICATION.

42. The inimical suggestion of some sceptics often takes this form: " Why should Electricity be the *only* agent in Nature possessing such a wide range of curative powers that it may almost be stamped as a panacea, a universal remedy, whereas up to the present time *various* curative results have only been obtained by the administration of a *variety* of drugs ?" In reply to this, it is only necessary to refer these sceptics to the fact, scientifically and generally received, that Electricity is a most powerful *stimulant* when applied in *one manner*, and exercises a most *soothing* influence when *differently* applied, owing to the fact that it causes decomposition and combination where no other chemical or therapeutical agent will produce any impression. Therefore, the list of diseases on which it exerts its healing powers, and their variety, as already given in extracts from various authors, is not to be wondered at.

Thus we see one and the *same* natural force possessed of a *variety* of curative powers, producing effects which, by the ordinary treatment, are obtained only from the application of a *variety of substances* in the form of drugs. But when we consider that the internal use of the latter very often throws noxious materials into the circulation, which are not always entirely eliminated from the system, either by the efforts of Nature or the administration of antidotes, it is not difficult to strike the balance between the *harmlessness* of Electricity as a powerful remedy compared with that of *drugs*.

INTRICATE CASES BEST TREATED BY A SCIENTIFIC MEDICAL ELECTRICIAN.

43. From the foregoing the inference may be drawn, that whilst, in certain intricate cases, the use of Electricity under the guidance of a scientific medical electrician may lead to the greatest curative effects, the application of the same means by the patient himself may possibly, in complicated cases, result in only partial cure, though *without any risk of injury*.

44. There are, however, hosts of ailments where even the haphazard application by the patient himself of PULVERMACHER'S Chains and Bands have had the most gratifying results, not only satisfying the patient, but utterly astounding the medical practitioner, whose skill had been baffled.

The object of this treatise being the enlightenment of all those interested in the progress of Medical Electricity, the following case may be quoted in confirmation of the foregoing statements:—

45. In August, 1866, when the cholera was raging in London, Miss Headley, a lady then living at 43 High Street, Kensington, called on me for the purpose of having her Chain-Battery (which was of the original construction) repaired. In the course of conversation, she spoke of the use which she was about to make of her Battery. She was afraid of the diarrhœa, to which she was greatly subject, and of which she had cured herself by the use of her Chain-Battery; and she wished to be prepared against an attack of the same disorder, which might reappear during the prevalence of the cholera. On her leaving my establishment, I was struck by the fact that Electricity, which up to that time I had known to be an excellent remedy in cases of constipation, should also be effective in a case of such an entirely opposite nature. Presuming that the mode of application adopted by the lady differed from that which was customary amongst medical electricians in cases of constipation, I corresponded with her, in order to learn the manner in which she had applied the Chain-Battery, and this is the answer I received:—

"43 High Street, Kensington, *August 4th*, 1866.

"Sir,—I am exceedingly sorry I did not receive your letter "until this evening, or I should have replied at once. In "replying to your first question, I must tell you that I do "not always use the conductors, but whenever I need them "I use instead two plates of zinc; one I place between the "shoulders, the other at the bottom of the spine. I let them "remain so about twenty minutes; then I take the one from "between the shoulders, and place it at the chest, and let it re-"main for two or three minutes, and I always find that checks "the diarrhœa; and no one can possibly have suffered more "from it than I have, for I had it before using galvanism to "an extent that made me so weak and nervous that I was "unable to sleep at night, but now I can sleep; and as for "appetite, I have an excellent one when I use your Battery. "With regard to your publishing my testimonial, I can have "no objection whatever; you are at perfect liberty to do so; "and I again repeat, I am sorry I could not answer your "letter before.—I am, Sir, yours faithfully,

"(Miss) J. HEADLEY."

The Testimonial referred to in this letter runs as follows:—

"KENSINGTON, *August 2nd*, 1866.

"SIR,—Having received so much benefit from your Chain "Battery, I think it might be of service to my fellow-crea-"tures if this could be made known. I have suffered "severely from diarrhœa, and have been under treatment at a "good many of the London Hospitals, amongst others the "Brompton Hospital for Consumption, where I was under "the care of Dr. Thompson for some time, but I did not "gain any strength from the medicine, and at last, having a "severe attack, I got so weak that I was unable to attend, "when I was persuaded to try your treatment. The effect is "quite marvellous. I find that by applying it for twenty "minutes it checks the diarrhœa, and, by repeating the same "for two or three days, I feel as well and as strong as I did "before the attack. It gives me new life and strength, and "is the only thing I have found do me good; and I am "thankful I found it out, and hope that others may benefit "by it.

"I am, Sir, yours gratefully,

"(Miss) J. HEADLEY."

Here we have a striking illustration of the importance of the mode of application when using Electricity as a therapeutical agent. In the case quoted, the electric action upon the nervous centres had brought about *tonicity*—that is to say, had given tone to the organs by the *continuous* current—whilst the peripheric or local application of the *intermittent* current, commonly used in cases of constipation, effected the cure by *stimulation* of the peristaltic action of the bowels. Many other cases have been observed in which *different modes of application* have effected cures of a *different* and even *opposite nature*. In order to promote the rational treatment of various diseases by means of his appliances, Mr. PULVERMACHER is preparing a Treatise* which, in a simplified form, embraces the whole subject of Medical Electricity in such a manner, as to render it comprehensible without tedious study of the science.

ELECTRICITY THE GREATEST MODIFIER OF NERVOUS ACTION.

46. Electricity being the *greatest modifier of nervous* action, will often most surprisingly effect a cure which the most scientific deductions will be unable to foretell, and which was the least expected by the patient. It is, therefore, incumbent on anyone who is interested in benefiting a patient, not to abandon the case before he has tried every possible mode of administration, remembering that the science of medical treatment by electricity is yet in its infancy. A true knowledge of this science will have to be acquired by future medical practitioners by a thorough scientific training, which an enlarged curriculum of studies will alone enable them to obtain.

INTERMITTENT AND CONTINUOUS CURRENTS; THEIR MODUS OPERANDI.

47. In consequence of the rapid extension of the application of Electricity for the cure of disease, it becomes

* GENERALISED MEDICAL ELECTRICITY: How, When, and by What Means best to Apply it Practically and Rationally. Profusely illustrated.

every day a matter of greater moment that the essential difference in the mode of action of the *continuous* and of the *intermittent* currents should receive due consideration in medical practice.

It is impossible to lay too much stress upon the fact, that the use of *magneto-electric* machines allows of a cure being effected only in a very limited class of diseases; whilst, on the other hand, the use of continuous voltaic currents proves daily more and more successful in all the following cases:—Firstly, where pain (not occasioned by any organic lesion) is the principal symptom; secondly, where the sluggish circulation of the blood is to be stimulated, and congestion counteracted; and thirdly, where the functional organs, acting under the immediate influence of the sympathetic nerve, require to be revivified, or respiration, digestion, assimilation, secretion, and excretion promoted.

The *intermittent currents* generated by induction coil machines, or even by Pulvermacher's Chain Batteries in conjunction with an interruptor, though producing powerful muscular contractions, are inefficient in polarising the liquids in the animal system, and influencing their chemical changes. The reparative action of the *continuous current* produces effects analogous to those of the natural currents of electricity in the body *when the latter is in a state of rest;* whereas the effect on the system of the *intermittent current* is analogous to that of the discharges of electricity evolved under the influence of *muscular exertion*, and thus tends to the *exhaustion* of vital force rather than to its *increase*.

In other words, intermittent currents affect the

organism in the same manner as exercise does, exhausting any local superabundance of vital force, and removing effete matter accumulated during the state of rest; whilst the continuous current affects the chemical changes within the system, and favours nutrition and the restoration of vital force, when the latter may have become deficient either through over-exertion or by various other influences.

It is easy to understand why, in local paralysis, and all affections in which the *motoric* nervous system is involved, the *intermittent* currents may exert a beneficial effect upon the paralysed muscles, by stimulating them to involuntary contractile action, notwithstanding the fact that these currents are chemically impotent in modifying the morbid condition of the tissues or secretions.

The chemical and physiological action of the continuous voltaic current is analogous to the peculiar and penetrating mode in which heat is generated by this current. The whole of the molecules in the fluid inserted in a voltaic circuit become, in fact, not only heated, but simultaneously polarised; becoming charged, to a certain extent, with condensed oxygen in the vicinity of the positive pole, and with condensed hydrogen in proximity to the negative pole—a fact well known to philosophers, and which may readily be proved by means of chemical reagents.

The liquids in the living organism, in the form of blood, &c., being more or less in a state of continual circulation, will, when brought under the influence of a continuous voltaic current, undergo in a similar manner this polarising action; the only difference being that fresh quantities of living fluid will successively pass within the sphere of action of the electric circuit. The liquids in motion being thus gradually charged with electrified and condensed oxygen and hydrogen (as gases in the nascent state), increased energy is induced in those chemical combinations and changes of the circulating fluids upon which the vital economy is dependent.

The peculiar action, chemical and calorific, to which we have referred, is possessed also by the *natural* continuous currents in the system, which constitute the main agency by which the normal functions of the animal economy are maintained; and this explains the generation of uniform warmth throughout the organisation during the processes of life.

These considerations sufficiently explain why the continuous voltaic current so often proves unmistakably beneficial, *in cases where the intermittent current has been tried in vain*; and fully justify the assertion, that no one who has made use *only* of an inductional electrifying machine, should imagine he has taken adequate steps to avail himself of the curative power of Galvanism.

There are many persons who believe that the portative Chain-Bands become inactive so soon as the liquid with which they are charged has evaporated, and that wearing them in their apparently dry state is merely a useless encumbrance. A simple test with the electroscope will suffice to prove that this is an erroneous impression. Even when the Band appears to be quite dry, it causes the repulsion of the gold leaves of this instrument when one pole is brought into contact with it, the other pole being held in the hand. But when the Chain-Band is slightly damp or humid, through being worn in contact with the skin, it produces a much greater divergence of the gold leaves—especially when the Chain has once passed through vinegar, in which case the acetate of zinc formed augments the effect caused by the access of oxygen to the negative element, which is secured by the peculiar construction of this miniature battery. Treatment by means of Pulvermacher's portable galvanic arrangements is, therefore, *not only more practicable*, but also more *efficacious* and rational, than the ordinary galvanic treatment; and the many successful cures which have been effected by the use of these flexible Batteries, in cases where the ordinary electro-therapeutic systems have failed, are easily explained by the facts already stated.

48. The most important practical feature, in connection with medical treatment by means of *continuous currents*, is the *direct* and *prolonged application*, without the intervention of conducting wires, realised through the perfect portability and electrical efficiency of my FLEXIBLE GALVANO-PILINE BATTERIES, wearable as a belt, &c. The *gentle* current which is *constantly* maintained by this combination influences the part affected in a manner *congenial to the working of the natural currents in the nervous system*, the reparative effects of which are sustained and strengthened. By this means warmth is restored in cases where its absence is a dangerous symptom, and vital action is *permanently* increased, an effect which ought not to be lost sight of in cases where this is requisite.

49. The importance of applying *a mild voltaic current* during a *prolonged period* is most strikingly exemplified by the action of a feeble current upon a piece of meat,

when the latter is placed in the circuit during a few days. The meat is preserved from putrefaction, and its chemical composition is also affected—the saline bodies being transferred to one of the poles, and the gelatinous matter to the other. Now, if the same quantity of electricity which, when acting for a lengthened period of time produces this result, were concentrated in an intense current of a few minutes' duration, it would produce the heating effect previously referred to, and result merely in a coagulation of the albumen and wasting of the meat. This difference of action in a *mild* current continued for a *prolonged* period of time, and an *intense* current *momentarily* applied—which is here illustrated by the effect produced upon *inanimate* matter—can but have its counterpart in the different modes of action proper to the two descriptions of currents in their effects upon the *living* organism.

50. In the electrotype we perceive another striking instance of the importance of a *slow* but *steady* electrolytic action of the galvanic current, *the precipitated metal being most coherent and uniform in texture when its deposition has been effected by means of a weak current, brought into play during a considerable period of time.*

RELATIONS OF ELECTRICITY WITH THE ELECTRO-PHYSIOLOGICAL WORKINGS OF THE ANIMAL SYSTEM.

51. From the peculiar mode of action exercised by electricity in a physical and chemical direction, it may be inferred that the *vital economy* of the animal frame, as already shown, is *strongly analogous to and intimately connected with electrical action.* This inference is fully borne out by a long and exhaus-

tive series of electro-physiological experiments carried out by Galvani, Aldini, Humboldt, Manteucci, Dubois-Reymond, Brown-Séquard, and Claude Bernard, who have published the results to the world. By these results it is shown, without any possibility of refutation, *that in the nerves*, and in the *muscles*, and even in the *blood*, currents of electricity are *continually circulating;* and by this circulation the *functions of life* are carried on and maintained, *movement* or *sensation* performed, and the *vital heat* generated and diffused throughout the system. We are thus able to understand that an abstraction of this native electricity, or the deviation of its normal and natural course, will occasion a disturbance in the health of the individual; and that, by artificially administering the electricity from *without*, the deficiency *within* the system may be compensated. In confirmation of this view, I could cite a long series of physiological experiments, which form part and parcel of the modern science of Electro-physiology, by which many of the phenomena of life hitherto obscured in mystery have been cleared up.

NERVO-ELECTRIC CURRENTS.

52. Aldini, nephew of Galvani, still further advanced those Galvano-electric investigations inaugurated by his illustrious relative. He passed a strong Electric current from the supra-orbital nerve down to the nerves of the heel of a man recently deprived of life, with the following results:—*The legs and feet moved rapidly, the eyes and closed, the mouth, cheeks, and all the features were opened brought into motion, and smiles and other varied expressions were successively depicted on the countenance.*

53. Brown-Séquard, an English physiologist of great distinction, took a living animal which had just been fed, and cut off the action of the nerves leading to the stomach. Digestion immediately stopped, and the animal began to show signs of starvation. On passing a galvanic current through the course of the severed nerves, the suspended action was *restored*, and the process of *digestion* resumed as regularly and completely as during the natural life of the animal.

54. Claude Bernard, the great French physiologist, also made a similar discovery. He divided the sympathetic nerves of the neck, when the corresponding muscles became greatly swollen and inflamed; but immediately he brought an electric current to circulate through the severed ends, both the *swelling* and *inflammatory* state *subsided*. The decisive experiment of Dubois-Reymond, a celebrated Prussian physiologist, demonstrated in the most direct way that the nerves and muscles generate and distribute in the system an electric force of great similarity to the continuous current evolved from a galvanic battery. By holding in the hands conductors, attached to the wires of a delicate galvanometer, the magnetic needle, on bending one arm, instantly deviated in *one direction*, and on bending the other, it as readily deviated in the *opposite direction;* thus showing that at every mechanical movement of the muscles *the body gives off an electric charge.*

55. The illustrious Humboldt combined a quantity of prepared frog's thighs together, so as to bring the nerves of one thigh into immediate contact with the muscle of the next thigh, and so on, forming in this way an animal Volta-electric pile, without the employment of any metal whatever. Immediately he closed the circuit, all the thighs

were thrown into violent jactitation—*thus manifesting in the animal organs an energetic galvano-electric current.*

56. The same philosopher, during his travels in America, communicated the most remarkable facts, illustrating the existence of Electricity in animal bodies. He says: "The Electric Fish, on being disturbed or excited to fear, discharges, as a means of protection, the most *terrible shocks of Electricity*, even of sufficient intensity to stun horses, and destroy the fishes upon which it feeds. In a short time, however, it expends all its force, and, like the overworked constitution, becomes totally prostrated; but after a certain interval it regains, by some vitalising process, all its former electrical energy."

57. If we consider that man becomes physically prostrated after violent rage, excessive grief, unreasonable muscular labour, or sudden mental emotions; that his constitution is severely affected with pain by every atmospheric disturbance; that the most startling waste of the whole system follows on disappointed affection, or sudden bereavements; and that by appalling fear his hair, by some electric repulsion, stands on end, and not unfrequently emits sparks clearly observable in a dark room,—then we cannot but admit that these conditions are effects arising from a cause common alike to all those phenomena represented in the above experiments. That is to say, as in the case of the Electric Fish, there is a modification of the *electric state in the body, or debility, caused by loss of Vital Electricity.**

It is easy, therefore, to comprehend, if the nerves play this all-important part in sustaining the body, that in the direst disease, either of the blood or of the nervous

* See Special Pamphlet.

system, the patient may, by surrounding himself with a net-work of electric fluid, arouse the dormant functions, and permanently restore himself to perfect health and strength.

KNOWLEDGE IS POWER, PROVED BY PROGRESS IN THE HEALING ART.

58. The proverb that "Knowledge is Power" has never been so truthfully applicable as in the case of medical progress during the last half century. The medical faculty, after wandering from one system to another during many centuries in a labyrinth of errors, have only advanced with the progress of positive science (Physiology, Natural Philosophy, and Chemistry), and are indebted to it for the present means whereby they combat those diseases hitherto considered incurable. *One of these means is secured by Pulvermacher's popular invention, not only to the medical profession, but to every sufferer.*

The author is fully aware that many persons not thoroughly acquainted with Galvanism and its wonderful properties, and even medical men who have employed Electricity by means of the old machines, will feel astonished, and perhaps incredulous, at the unlooked-for therapeutic effects of *weak* but *continuous currents*, as produced by these GALVANIC CHAIN-BANDS. Truly, as Wesley observes,* "How many lives could be saved by this unparalleled remedy. And yet with what vehemence has it been opposed! Sometimes by treating it with contempt, and as if it were of little or no use; sometimes by arguments, such as they were, and sometimes by such cautions against its ill effects as made thousands afraid to meddle with it."

* THE DESIDERATUM, OR ELECTRICITY MADE PLAIN AND USEFUL. Re-edited by Baillière, Tindall, and Cox, London, 1871.

59. It is requisite that a few words should be said relative to the source from whence the Electricity in the system is derived. With every breath of air which our lungs inhale, the venous blood is not only oxydised and transformed into arterial blood, but it is also charged with Electricity, produced by the condensation of the air, which takes place by the pressure through the bronchial tubes of the lungs, during the act of exhalation. That this is actually the case is proved by the experiment of Dr. Kincke, of Berlin, showing that currents of electricity are engendered by pressing an aqueous liquid or damp air through a membrane of bladder, or silk, or even through a diaphragm of sulphur in a powdered state; the greatest quantity of electricity, equal to that generated by a Daniell element, being yielded by the latter.

THERMO-ELECTRICITY, ITS INFLUENCE ON VITAL ELECTRICITY.

60. Amongst the many varied sources which generate Electricity, difference of temperature holds a prominent place, as it is most intimately connected with the functions of life. This Electricity, deriving its power from heat, is named Thermo-electricity. To illustrate this, take a piece of wire, and bring its ends in connection with those of the galvanometer; heat one of these ends by the flame of a spirit-lamp, and leave the other cold, or even lower its temperature by artificial means; an electric current will be produced, which will show itself by the deflection of the magnetic needle of the galvanometer. What takes place in metallic wires owing to difference of temperature occurs also, more or less, in other material, liquid or solid, in accordance with its con-

ductibility and its unalterable nature, by the influence of preponderating heat and cold.

To the electro-generating properties of difference of temperature may be attributed the beneficial effects observed in certain cases, in which from time immemorial central applications of cold (ice-bag) to the head and spine have been beneficially used, while warmth applied to the lower extremities produced similar results; thus furnishing a current of electricity flowing in that direction which appertains to the animal frame in the normal state of health. The morbid symptoms observed after taking cold are, therefore, reasonably attributable to the dispersion of the normal electricity, and generation of electrical currents in an *adverse direction* by the *perispherical cold;* as is the case when, after energetic muscular exercise, which produces perspiration, a draught is allowed to cool the heated body or the extremities externally, by evaporating the moisture thereby produced.

61. Ever since the first means for the artificial production of Electricity were introduced, its medical application has been eagerly resorted to as often as a new progress in the electrical science, and a fresh advance in the construction of the apparatus for generating and applying it, have been made, and this not without good reasons. When Electricity is applied as a remedy, no foreign matter or substance is thrown into the circulation, as is the case with drugs; it having been shown that Electricity is a pre-eminent force in nature, initiatory to a variety of chemical and physical changes accompanying the processes of life. It is also capable of influencing and regulating the latter to such an extent as to aid Nature in her efforts to re-establish the

balance of health. By so doing it assists any other remedies, no matter what their nature may be—drugs or diet, or any other natural agent, pure air, &c.—calculated to exercise a reasonable beneficial influence on the patient. Therefore, a belief that Electricity is not applicable simultaneously with other remedies must be dismissed as illogical and unreasonable.

62. There is a subject of great interest worthy of the attention of all those who share the opinion that "prevention is better than cure." Galvanic electricity not only rivals any known preventive means, but excels them by far in its efficacy for preserving health; it protects the system against pernicious influences from without, which tend to disturb the normal (healthy) functions of life. If bathing, exercise in fresh air, careful attention to clothing in regard to the changes of the atmospheric temperature, diet regulated with reference to the choice, quantity, and time of eating and drinking, are acknowledged as means for the preservation of health, so much the more should Electricity be considered as a preventive: and in this direction the modern means of application, as furnished by these various inventions, are assuredly invaluable. How many derangements occurring during the period of teething might be prevented if mothers were but aware of the fact, that the popular use of amber beads, worn with advantage, from time immemorial, as necklaces for children, is based on the Electricity resulting from friction of the amber caused by the restless movements of the child. The Galvanic Electricity *supplied* by Voltaic Necklaces advantageously replaces the frictional Electricity of the amber beads.

Taking into account that mortality is, according to statistical returns, greatest during the period of infancy, and that a mild supply of Electricity reacts beneficially on vitality during childhood, in a degree exceeding that in adults, it is incumbent upon parents to afford their children that protection against the derangements referred to. While treating on the use of Electricity as a *preventive*, the subject of the preservation of the healthy condition of the teeth, as well as the gums, is not of less importance than the cure of a disordered stomach. To carry out these ends, I have used my best efforts to enable those who see the import of gentle electrical action in the healthy nutrition of gums and teeth, and its counteracting power against putrid formations (the ferment of many a decayed tooth), to cause a galvanic current to act upon the mouth as often as the teeth are brushed. For this purpose I have constructed a tooth-brush, which will convey to the mouth a gentle current of continuous Electricity, thus securing at a small outlay the above-mentioned advantages.

63. But to conclude. The progress of Medical Electricity is intimately associated with that in physics, chemistry, and electro-physiology. These exact sciences, based as they are on verified facts, discovered through the indefatigable researches of hosts of scientific men, preclude all arbitrary interpretations, and are therefore the most reliable guides for further progress.

Electricity is now only in its infancy. Its future is illimitable, and wonderful discoveries await the patient and laborious student. Who could have thought that the accidental contraction of the muscles of a frog would ever have paved the way to such brilliant results as have already appeared ? No fairy dream could ever surpass the wonders of the present age ; and even the lively imagination of Lord Lytton, in "The Coming Race," could not foretell the vastness of future discoveries, to which the present, though wonderful, are but as the embryo. While heat now carries us with the swiftness of a gunshot whithersoever we will, the heralding electric messenger is used as a safeguard against the otherwise destructive speed of the locomotive engine, Electricity likewise conveys our thoughts to dear and distant friends, by means of a tiny thread of wire dangling in the air or submerged in the ocean. That a simple ribbon of metallic texture, of zinc and copper wire, should minister so powerfully to us in our affliction, is the result of discoveries which have cost the author persevering labour and self-abnegation during a lifetime ; but have at length enabled him to give to the world, in the cheapest and most convenient form, not a panacea, but a medium of healing or relief for many of the "ills that flesh is heir to." The best medical works on this and the other side of the Atlantic, abundantly testify the cures the "Pulvermacher" appliances have effected ; even in the case of patients who applied it themselves, and who, moved by a feeling of thankfulness, communicated their testimony, which is embodied in a pamphlet, entitled "Galvanism, Nature's Chief Restorer of Impaired Vital Energy."

AN HISTORIC SKETCH OF PULVERMACHER'S GALVANIC INVENTIONS.

Most of the friends of inventive progress, who have taken an interest in my various electrical appliances for the cure of disease, have been desirous to know something of the history of my inventions.

In anticipation of the same token of interest on the part of others, I here venture to furnish a brief outline of my earlier history, and of the endeavours which have occupied the best portion of my lifetime.

In the years 1843 and '44, prepared by studies pursued during my leisure hours, I regularly followed the lectures on Natural Philosophy by Professor Hessler, then of the University of Prague, and subsequently of the Polytechnic School in Vienna.

The practical applications of Electricity, which were then being introduced in the direction of telegraphy, &c., &c., possessed the greatest attraction for my intuitively inventive disposition. The plan of a new electro-magnetic engine for producing motive power, which I laid before Dr. Hessler, Professor of Physics, and afterwards before his successor, Professor Petrina, and also Professor Kreil, met with their hearty approval and encouragement, as may be inferred from the following extracts from two of their testimonials :—

"Prague, 19th September, 1845.

"The undersigned hereby certifies that he had the oppor-
" tunity during the last year for frequent discussions with
" Mr. Pulvermacher on the various branches of physics, and
" especially on Mechanics and Electricity, and that he has
" found in him a very able and diligent "autodidact," who by
" his excellent powers of conception, by the exuberance of
" original ideas, and love for these sciences, could, if more

" favourably situated, render useful services by the studies " already begun, not only to science, but through his mecha- " nical dexterity, also to practice.

" DE. FRANZ PETRINA,
" Royal Imperial Professor of Physics and Practical
" Mathematics of the University of Prague."

" Prague, 27th May, 1846.

" The undersigned hereby certifies that Mr. Pulvermacher " has made profound studies on the nature of the Electro- " Magnetic currents and their application to motive power, " and by several experiments made in his presence, and that " of other scientific persons, has indisputably proved that he " is able, by new electro processes, to increase the Electro- " Magnetic power to a degree never yet attained.

" KARL KREIL,
" Prof. of Astronomy at the University of Prague, and Director
" Astronomic and Magneto-Metric Observatory."

Through the patronage of these scientific men I was enabled to proceed to Vienna, in order to be nearer the more centralised resources for carrying out this and some other inventions in connection with Electricity.

Settled in the German metropolis 1846-47, I was so fortunate as to secure the kindly interest of Professors v. ETTINGSHAUSEN and v. BAUMGARTEN. I here append a letter from the former :—

" Vienna, 21st June, 1846.

" Esteemed Doctor,—Mr. Pulvermacher has submitted to " me a plan for the construction of an Electro-Magnetic " Machine invented by him; although I should not vouch " positively for the practical results of a machine executed in " this manner, as only a model carried out of moderate size " can show the proportion of materials consumed, and the " power produced, yet the novelty of ideas which may even " be applied in other directions, and above all, the soundness " and manifold knowledge in natural philosophy which Mr. " Pulvermacher has acquired, in his limited circumstances " of life, by devoting every leisure moment to reading and "thought, have inspired me with the deepest interest in "the matter. . . . . . .

"Should you be in the position to draw the attention of "men, in the circle of your acquaintance, who are willing to "assist honest endeavours to pursue that branch of scientific "and practical activity for which he appears to be destined, "you would certainly devote your trouble to a good purpose.

"Excuse, worthy Doctor, the liberty which I here take; "it is solely founded upon the high esteem with which I "subscribe myself,—Yours faithfully,

"A. VON ETTINGSHAUSEN,

"Royal Imperial Councillor of State and Prof. of Physics."

"To DR. LUDWIG AUGUST FRANKEL."

Through their influence I obtained substantial assistance from the late Baron SALOMON V. ROTHSCHILD, towards the construction of models of my inventions, and obtaining letters patent in England and elsewhere.

At that period a chain made of zinc and copper, and falsely represented as being electric, was largely advertised in Vienna by one "Goldberger" as adapted for medical use; but it was universally condemned by the Faculty as a deception. In order to prove by comparison the faulty construction and spurious nature of this appliance, I was urged by several of my medical friends to invent a *bona-fide* Electric Chain; which, although at that time only a rough model, produced such manifest electric phenomena as to interest many scientific men in its favour, and showed by contradistinction the futility of the "Goldberger" chain.* I was fain to yield to necessity, and to confine myself to the working of the patent relating to my original invention in Chain-Batteries. This I did the more readily in consequence of the encouragement and testimony which I received from

* Vide Extracts from "German Clinics," p. 80.

Dr. OPPOLZER, Clinical Professor, of Vienna, the late Drs. GOLDING BIRD and PEREIRA, of London, and several of the Members of the Academy of Medicine and Science in Paris. The following are extracts :—

"Vienna, 1849.

"I hereby certify that Pulvermacher's Galvanic Chain may "be employed with great success in cases where electricity "is employed in Rheumatism, Paralysis, Debility, &c.

"Professor J. OPPOLZER,
"Chief Physician of the Imperial Hospital at Vienna, and
"Physician to his Majesty the Emperor of Austria."

"48 Russell Square, Aug. 29, 1851.

"The ingenious modification of Volta's Pile, invented by "Mr. Pulvermacher, was placed in my hands several months "ago, and I have had the opportunity of testing its value. "This apparatus is capable of producing all the Physiological "effects of the well known Galvanic Battery, each link of "the Chain corresponding to a cell of the latter very cum-"brous and (for medical purposes) inconvenient machine. "It is easily excited, and its power is very persistent. With "careful management it is not likely to get out of order.

"We have in this ingenious invention that which has "long been a *desideratum*, viz.:—an apparatus of the smallest "possible bulk capable of evolving a *continuous or uninter-"rupted* current of Electricity of moderate tension and "*always in one direction*, without the expense, bulk, and "great inconvenience of the Cruikshank Trough or the "other cell arrangements. I can hardly recommend Mr. "Pulvermacher's invention too strongly to the notice of my "medical brethren.

"GOLDING BIRD, A.M., M.D., F.R.S.,
"Fellow of the Royal College of Physicians, Physician to and
"Lecturer on Therapeutics at Guy's Hospital."

"Finsbury Square, Sept. 9, 1851.

"I have great pleasure in stating, that I consider "Pulvermacher's Patent Portable Hydro-Electric Chain "Batteries to be a very convenient and effective form of "Voltaic Apparatus for medical purposes.

"JONATH. PEREIRA, M.D., F.R.S., F.R.C.P.,
"Phys. to Lond. Hosp."

Extract from a letter from Dr. DUBOIS to M. BERARD, "*doyen de la Faculté*:"

"June 8, 1850.

" . . . . Be pleased to examine and experiment for " yourself the Voltaic Chains which are shown to you by " the bearer, J. L. Pulvermacher. I have seen nothing more " ingenious, more portable, or more powerful."

Letter from Dr. DUBOIS, Perpetual Secretary of the Academy of Medicine of Paris, to Mr. PULVERMACHER:

" April 3, 1851.

" MONSIEUR,—I am happy to inform you that the com- " mission appointed by the 'Academie' to examine your " Voltaic Chains made their report, through M. Soubeyran, " on Tuesday last.

" The 'Academie,' after having heard the reading of the " report, have desired me to express you their thanks for " your important communication.

Believe me to be, yours faithfully,

(Signed) " DUBOIS."

(*Short Extract of this Report, see page* 30.)

Letter written by Dr. Koreff, Privy Councillor to H.M. the King of Prussia, and Knight of several Orders, to M. Foucault, Professor of Physics at Paris:

" My illustrious friend, allow me to introduce to you " my fellow-countryman, Mr. PULVERMACHER, who has made " some important discoveries in Galvanism, which he has " already communicated to the Institute, from which he has " received reports through MM. POUILLET and BECQUEREL.

" Amongst other inventions he has discovered a real Gal- " vanic Chain, which for its therapeutic action offers great " advantages. He has also made other galvanic discoveries " of great importance, for which he has already taken out " patents in England and France. The object of these is to " regulate motive power and prevent the waste so much to " be regretted.

" He is a very remarkable man; originally a working " jeweller, who, urged by an indefatigable instinct " towards chemistry and physics, learned these sciences " whilst struggling for existence. He is now at the head of "extensive mechanical works in Vienna, and has become "the *protégé* of the most eminent scientific men, and has "already met with great success in England. He is a man of "great ability, through which, and noble perseverance in a "laborious career, he has acquired the right of being well "received by men of genius. It is for this reason that I "address you.—Yours sincerely, KOREFF, M.D."

Following the advice of friends, I settled in France in the year 1850, and in England in 1859, for the purpose of carrying into effect the Medico-Electrical part of my patents, postponing at a sacrifice the working of the other portions relating to electro-magnetic engines, telegraphy, &c. (Patent, Nos. 12,899 and 13,933, *Old Law*).

Thus at first almost involuntarily drawn in the direction of MEDICAL ELECTRICITY, I soon perceived that, by reason of the services which this science is capable of rendering to humanity, there was ample scope for my activity in this sphere, and opportunity for a beneficial and useful mission. Unfortunately, my want of acquaintance at that period with the French and English languages, caused a long series of years to be expended in struggling with great disadvantages. I was compelled to launch my first Electrical Chain through a sole agent, "Mr. Meinig," then of Leadenhall Street, whose only idea, I subsequently discovered, was lucre; and who, failing to find sufficient support from the medical profession in general, addressed himself to the public, and with great success. His manœuvres to despoil me of my patent rights in France and England resulted in a lawsuit, the nature of which may be gathered from the address of the celebrated M. Cremieux to the Imperial Court, Paris. (See p. 80.)

The verdict, obtained first in Paris, and afterwards confirmed in London, served only to reinstate me in my patent rights, for, through my adversary decamping, the damages awarded, viz., £10,000, remained *non est.* Being obliged to look after my manufactory in Paris, I was compelled to take a partner, to whom I entrusted the management of my London business; he however, didnot improve my position materially, not being possessed of the necessary attainments for the working of my *specialité.* Tenacious perseverance, however, at length enabled me to surmount the many difficulties thrown in my way.

In order to turn my long experience and knowledge of sciences connected with Medical Electricity to good account, for the progress of Electro-Therapeutics, I again set to work in 1862 to carry out improvements in my Flexible Voltaic Batteries, in the form of *Chain-Bands*, and also to devise different appliances for the more simple and practical utilization of the electrical agency in the treatment of certain classes of disease. The successful accomplishment of this task is proved by the testimonial, signed by Sir CHARLES LOCOCK, Sir HENRY HOLLAND, Sir WILLIAM FERGUSSON, Sir RANALD MARTIN, and others of the *élite* of the medical profession. The following is a quotation :—

"We have much pleasure in testifying that Mr. J. L. "Pulvermacher's recent improvements in his Voltaic Bat- "teries and Galvanic Appliances for medical purposes are of "great importance to scientific medicine, and that he is "entitled to the consideration and support of everyone "disposed to further the advancement of real and useful "progress."

In the year 1868 a certain class of greedy speculators,

in order to serve their quackish ends, were advertising sham galvanic contrivances as a bait for the unwary. By them the value of the above testimonial was so well understood, that no sooner did I publish it than they at once sought to create confusion, by endeavouring to make the public believe that this patronage applied to their sham productions. This manœuvre I succeeded in arresting by suing for and obtaining an injunction from the Court of Chancery (August, 1869). See "Sincere Voice of Warning, &c.," by J. L. Pulvermacher, 1869.

Having from that time perseveringly continued this task of improvement, I have finally succeeded in bringing to a very high degree of perfection my system for the *steady* application of *intermittent* and *continuous* currents, by means of new inventions, already described.

In pursuing the improvements which have so greatly facilitated the self-application of voltaic appliances, I at the same time ever bore in mind those which were calculated to render my system completely adapted for extended use by the medical profession. From the period when, at the suggestion of my medical friends in Vienna, I devised and carried into effect the first genuine voltaic chain (see p. 68), it has always been my conviction that this invention is destined, sooner or later, to become a powerful auxiliary remedy in the hands of general medical practitioners—for whom, indeed, it was originally intended. A question may here occur to the reader which has frequently been addressed to me, both by persons who were sceptical as to the degree of efficacy possessed by my voltaic appliances, and also by those who, having experienced their effects, were unable to account for the neglect they have incurred :

"How is it that this invention has made comparatively so little way amongst the generality of the medical profession, although more than a quarter of a century has elapsed since it was first brought to light under such favourable auspices? And how is it that an invention of this character should need to be placed before the public by means of an extensive system of advertising?"

The true answer to this question does not, as I believe, involve any selfish or unworthy motive on the part of the medical profession. Having all my life been on fraternal terms with many medical students and practitioners, both in England and on the Continent, I have had good opportunities of observing that medical men, in general, find a true gratification in serving their fellow-creatures, and are interested in their relief, not only by the powerful incentive of professional reputation, but through the yet higher motive of moral satisfaction. The real cause of their ignoring the progress of the healing art in the direction of electricity appears to be, that the greater number of medical men have had no opportunity, whilst students at college, of acquiring that indispensable knowledge of physics which is intimately connected with, and involved in, the study of Electro-Therapeutics and Electro-Physiology. These sciences were not, in fact, included in the curriculum of their studies; and, once embarked in practice, they have neither the leisure nor the experimental means for remedying the deficiency. Hence, fearing that medical electricity, in their hands, might probably be ineffective, they think it prudent to refrain altogether from its use. Had this class of practitioners been sufficiently acquainted with physical science to perceive how readily they might avail themselves of the curative powers of electricity, and how perfectly they could control its effects, by means of my medico-electric inventions, I have no doubt but that they would long ago have adopted them as auxiliaries to their usual modes of treatment. Considering the favour with which these inventions have been received by the highest scientific authorities amongst physicists and medical men in every country, it has been with many a matter of surprise as well as regret, that I should still be compelled to have recourse to advertising, instead of receiving freely from medical practitioners that extended patronage which the inventions deserve. The cause for their surprise may be removed; that for regret still remains. It is hardly to be doubted that, if patients generally had been enabled to apply the voltaic curatives by the prescription and under the direction of their usual medical attendants, the number of cures effected would have been vastly increased, and would have redounded in greater measure to the advantage of practitioners and patients, as well as to that of the invention itself. And, moreover, in this case the public would have been relieved by medical practitioners, their natural protectors in a case of this kind, from the necessity for distinguishing between these inventions and the spurious appliances which are so frequently brought forward by ignorant adventurers, speculating on those unversed in such matters.

To M. VICTOR MEUNIER, *Editor of the "Amides Sciences."*

DEAR SIR,— I am, as you well know, one of those who revolt at injustice and feel stung at ingratitude. Now, as there is nothing more brutally unjust than ignorance, nor so ungrateful as that part of the public called the *masses,* I never want for occasions to excite my bile.

One inventor who is not appreciated, is set down at once as a fool or a madman. Another on the contrary, has seen his enterprise crowned with success, either through his happier choice of means for its execution, or because he had the good fortune to satisfy more immediate wants, and he instantly becomes the butt of envy: piratical imitators assail his fortune, and at the same time endeavour to depreciate his merits and strip him of his share of glory, that powerful incentive of the inventor, by stigmatising his work by the name of quackery.

These sad reflections occurred to me with redoubled force the other day, when I was conversing with an excellent doctor of my acquaintance, on the indisputable progress made by electricity in the practice of therapeutics. It is very clear that this universal principle, perhaps the sole principle of motion, heat, and light, all which may perhaps be only one under this threefold form, it is, I say, evident that this most subtle agent cannot be without effect on the animal economy of which it forms part, and that whatever changes the equilibrium, produces a modification. Now, to modify the animal economy when in a state of health, is to produce a morbid state; so, to modify it when in a morbid state, affords at least a chance of producing a return to health.

As a case in point, we began to talk of Pulvermacher's Chains, which are advertised everywhere. My friend did not hesitate to avow that he had little confidence in such apparatus, and greatly doubted their efficiency. In his mind, Pulvermacher was only a clever charlatan, much better versed in puffing than in science. Besides, was there really such a man? He was even sceptical enough to question that. I must confess that I was still influenced by the good doctor's doubts, when I called at Mr. Pulvermacher's establishment, for the purpose of seeing and testing his Chains. I now send you the result of my investigations, with the conviction that many scientific and medical men will have to thank the

"*Ami des Sciences*" for directing their attention to one of the most convenient sources of electricity, for all experiments of short duration.

The extreme courtesy with which Mr. Pulvermacher supplied me the means of making experiments with his Hydro-electric Chain, and measuring its various degrees of strength, allows me to describe it in detail.

It is evident that the Pulvermacher Chain Battery constitutes, in a small bulk, at once a voltaic pile and an electromotor, by the help of which all the experiments peculiar to this kind of electricity may be performed. . . . .

What most surprised me, I confess, was the very extraordinary persistence of these phenomena, which I reproduced several times, without any apparent difference in the intensity of the effects, during nearly an hour.

When the effects are no longer manifested, it is only necessary to dip the Chain for an instant in vinegar, and the action is renewed instantaneously after each immersion. *The Chain requires no arranging or cleaning, either before or after its use. The complete apparatus is of the size and shape of an ordinary pocket-book.* . .

Mr. Pulvermacher, we must not omit to state, is one of those men with whom you can hardly come in contact without learning something.

A distinguished experimentalist, a clever manipulator, an ingenious mechanist, and an inventive mind, he has been engaged nearly all his life in the study of practical electricity, and I have seen at his establishment several instruments and apparatus, both modelled and in course of construction, which cannot fail to do honour to his name, and procure him well-merited eulogiums.

Not to mention a telegraph, with a keyboard, and an apparatus to regularise the intensity, hitherto so variable, of the electric light, he is now making a voltaic pile for a single fluid, the constancy of which (theoretically speaking) must be indefinite, or lasting so long that it may be considered so. I have thought it a bounden duty to render justice to Mr. Pulvermacher, and at the same time to promote the interest of experimentalists, by calling their attention to the Hydro-electric Chains of that clever inventor.

GAUGAIN, Physicist and Engineer, &c.

From "THE SCIENTIFIC AND LITERARY REVIEW," April, 1873.

"It is a singular fact that the first useful application of Electricity was that of restoring health. The improved means of practically applying this subtle power is an important item in the progress of electro-therapeutics, and in the extensive and ever-growing literature explaining the various medico-galvanic appliances in use, we find, side by side with other scientific apparatus, Mr. Pulvermacher's various inventions of Voltaic Chain Bands, Batteries, &c., frequently treated upon in terms flattering to the inventor. . . . . . . As we have above seen, the success, both scientific and general, is owing to *great simplicity*, coupled with great electric efficiency; and this has, therefore, induced various other persons to put forward contrivances professedly possessing similar powers and virtues; but these persons, either from ignorance of the scientific cause of the efficiency of Mr. Pulvermacher's appliances, or else dreading the penalties attending the infringement of his patent rights, claim to have found the secret of producing portable electric and magnetic contrivances *without the use of an exciting liquid* and *without magnets*, thus *endeavouring to mislead the uninformed*. . . . The invention of Mr. Pulvermacher, we find, has been described and favourably commented upon in numerous scientific works."

Extract from a translation of the speech of M. CREMIEUX,—Ex-Minister of the French Republic, the eloquent advocate of M. PULVERMACHER, in the action brought by him against a M. Meinig, for fraudulent breach of trust—delivered before the Imperial Court of Appeal, in Paris. The action in question resulted in a judgment for the plaintiff, by which the dishonest infringer of M. PULVERMACHER'S patent rights was condemned in DAMAGES amounting to 200,000 fr., in addition to 50,000 fr. awarded by a previous conviction.—*See "Le Droit," 4th of February*, 1855.

"The facts of the case," said M. CREMIEUX, in his

memorable address, are as follows: PULVERMACHER is a German, born in Breslau, of a respectable family; gifted with a rare degree of intelligence, he is one of those speculative minds often to be met with in his country, following with ardour an idea which they believe to be useful to science, and sacrificing everything to obtain the honourable name which is the object of their ambition; if their idea happens to be largely carried out in practice, fortune may accompany this honour. Moreover, laborious as a *savant* living in his laboratory, unacquainted either with the world or with its trickery, a man of antique mould, easily duped whilst walking hand in hand with Meinig; such is Pulvermacher.

"Let me now explain to you, gentlemen, what is the invention of PULVERMACHER.

"*Electricity*, applicable in medical science—in that great science which protects human life—has for a long time past been the subject of most extended researches. Pulvermacher, perhaps, has effected a great step; of this you will be able to judge for yourselves, if you will for a few moments grant me your kind attention. The subject of which I am about to speak is historical rather than scientific.

"Those amongst us who were born at the close of the last century, or at the commencement of the present one, may remember the electrical experiments which took place in their youth. There was a large machine, in the centre of which was a mechanical arrangement for the purpose of communicating movement to a glass disc, a jar which was charged with electricity, a chain formed by a certain number of spectators holding each other's hands: the physicist held the glass jar, and the first person of the chain approached his finger to it, when the electricity produced a shock, a commotion experienced by each one of the persons forming the chain. . . . This shock, which might be rendered more or less violent, was applied in medicine against those terrible diseases which deprive the muscles of movement, of life,—to paralysis of the body. But the hopes which had been conceived were far from being realised. You may now find these machines in the cabinets of physicists—an object of curiosity, a memento of the progress of art.

"I briefly pass on to the machines which preceded the discovery of Pulvermacher.

"The electric shock is produced only at the moment when electricity enters the body; so soon as tho current is established and passes through the organism, chemical or physiological effects may take place, but not those producing motion. Improvement was to be sought by increasing the rapidity of succession, by multiplying, in fact, the shocks. A great English philosopher, Faraday, carried this improvement into effect by means of a machine termed the *Inductional Machine.* Here is one which I bring before the notice of the Court, as being necessary in order to enable us to understand the invention to which I havo particularly to refer.

"Such, gentlemen, with the exception of some improvements which are of no vital import, was in 1849 the state of science in relation to the interrupted electric current.

"Allow me also to say a few words on the subject of the continuous electric current. It is applicable in *nervous diseases,* in those where *pain* is a prominent symptom, and in rheumatism: these it cures or relieves.

"Thus it was, that, profiting by tho results of science, our most distinguished medical men succeeded, by means of electricity, in curing, or at least in alleviating, the worst disorders.

"The continuous electric current, like the interrupted current, is generated by the combination of two metals, and with the aid of an acid.

"I have caused to be brought beforo you the battery which at the present day is employed in the application of electricity to the human frame. It is here in this large chest; it is composed of twenty elements. Each element is constituted of a plate of zinc and a plate of copper.

"One word more. The electric current is developed only at the surface of the metals of the battery; therefore the more elements there are in a battery, the greater the intensity of electricity supplied, but the more cumbrous and costly also is the apparatus. From this battery of twenty elements you may judge what must be a battery of sixty elements! how difficult must it be to carry from place to place! how expensive it must be in price! This double drawback is greater even than in the inductional machine.

"What, then, is required in order to constitute a real improvement? If a single machine can be made to furnish both the *interrupted* and the *continuous* electrical current; if, contained within a limited space, it can yet supply a large quantity of electricity; if it is *portable* and *not costly*; if it can be set in action by the simplest means, and without producing any odour; if it can be applied not only by its extremities, but throughout its whole extent, so as to *surround the body of the patient*, to be removed only by order of the medical man,—is it not the case, gentlemen (putting aside all science, and judging only by common sense and reason), is it not evident that if such a discovery has been made, the results which may be expected from it are immense?

"The discovery of Pulvermacher realises all the conditions which I have just pointed out.

"This, then, is his invention. When the interrupted current is required, we need no longer use these large batteries to be placed inside, as in the case of the inductional machine. To obtain the continuous current, we no longer need these enormous batteries, with their mephitic exhalations, contained in this immense chest. A chain which you take in your hand, and of which you will readily understand the mechanical arrangement."

(*Here follows a description of the Chain Bands.*)

. . . . .

"Thus from an instrument contained in the smallest space may be obtained the largest quantity of electricity; this fact suffices to render it an excellent and portable apparatus! you can see for yourselves the size of the case which contains it. If you ask one of our fashionables for his cigar-case, it will hardly be smaller than this, which contains the whole invention. Thus the price sinks into insignificance.

"And this, gentlemen, is the ingenious system adopted by the inventor.

"These chains are so constituted as readily to envelope the portion of the body affected. The medical man adjusts one, two, or several of them; they remain in position as long as he may think fit. So much for the application of the continuous current in diseases attended by pain.

"Let us turn to the case where shocks, that is to say, the interrupted current, are required.

"At the period of his arrival in England, Pulvermacher

had constructed a little instrument in glass, which was to be placed between the two chains. This instrument, termed a *contact breaker*, produced an interrupted current by means of the vibration of a spiral spring. But, nearly three years ago, another ingenious invention for the same purpose, which was received with universal approbation, allowed of the most extended application of the interrupted current."

After a long and brilliant exposition of the legal question, M. Crémieux concluded as follows :—

"Gentlemen, I have now ended ; the fraud which has been shown, with all its machinations, and in all its audacity, the fraudulent intent manifested throughout, and clearly evidenced by manifold facts, calls for an exemplary punishment. The penal clause has been inserted in the contract ; you will apply it without hesitation. Pulvermacher will not in vain have counted upon your justice, and your sentence against the unworthy speculator, constituting a just expiation for his odious conduct, will make legitimate amends to the inventor, allowing him without fear to devote himself to fresh investigations. Justice, in punishing fraud, will upraise science."

Extract from an article in the GERMAN CLINICS *of March* 13, 1854, *on the Chains of Goldberger and Pulvermacher, the principles on which the latter are based, and the difference resulting from the comparative experiments of the two Chains, by Dr. Kuchenmeister, of Zittau, Saxony, in conjunction with Drs. Jahn, Dietzel, and Schmidt, professors of physics and mechanics to the Gymnasium and School of Arts and Trades at Zittau.*

We have examined Pulvermacher's chains, and we find the laws of galvanism manifested in their construction in the clearest and most evident manner.

2. The strength of the current is produced by several causes.

In the first place, the use of a wet conductor (acetic acid), a chemical excitant of the skin and the chain at the same time ; next, because the apparatus acts of itself, without depending in any way on the greater or less sensibility of the skin of the person wearing it.

If we confine our attention especially to the law (of galvanism) mentioned below, we shall also see the change of the current effected by the ingenious little spiral in the glass cylinder (Pulvermacher's "Interruptor"). By shaking the glass cylinder the spiral wire is set in motion, and consequently the flow of electricity is disturbed. It must be owned that the low price of these chains would hardly justify us in hoping for so simple an arrangement, and yet it is on this very point that we hope to see a still greater improvement of the apparatus effected by its clever inventor.

It would be quite useless to attempt producing any of these effects with Goldberger's chain. . .

I should be highly gratified if these lines contribute to make Pulvermacher's chains known to the public. The best means of acting against charlatanism would be to deliver lectures on galvanism in provincial towns, and show by actual experiments the vast difference between this Chain and other pretended electric instruments.

The Electric Chains of Mr. Pulvermacher combine great convenience of form with a most remarkable degree of electrical power, and they certainly deserve to be recommended in preference to all others for medical use.

MULLER, M.D.

ACADEMY OF MEDICINE, NEW YORK.

Extract from the Records of the Committee on Chemistry and Pharmacy to the above Academy, at the Meeting on the 1st of December, 1852.—Committee: Drs. MACNEVEN, SAYRE, GARRISH, and TAYLOR.

"In regard to the electro-voltaic apparatus of Mr. Pulvermacher, for medical purposes, every facility was afforded the Committee of testing its applicability. This instrument was found to be essentially an ingenious modification of the original voltaic pile: it consists of a Chain composed of a series of zinc and copper wires, well arranged; by simply moistening this Chain with vinegar, or acetic acid, it is prepared for action.

"A galvanic instrument of this form is well calculated for medical purposes, from being so readily available when required. In examining it, in reference to its advantages for ready application, the power, intensity, and permanency of action it was capable of manifesting, were also duly considered.

"The Chains exhibited to the Committee were of two sizes, with powers proportionate to their respective dimensions. The smallest sized Chain is intended where moderate power only is requisite; the larger Chain is intended to supply a power equal to any demand which would probably be made for medical purposes. One of the latter size, after having been once thoroughly moistened with vinegar, and left exposed to the air, was found to be capable of deflecting the galvanometer for two hours and a half. In the ordinary use of the instrument, however, the permanency of its action must be maintained by moistening the Chain from time to time, say about once in fifteen or twenty minutes, with vinegar. The number of elements may be readily increased by connecting two or more Chains together, and the intensity of the galvanic current proportionately increased. Four Chains, of sixty elements each, having been thus connected, the intensity of the electric current was found to be augmented to a degree too painful to endure.

"*On the facts above stated, the Committee is of opinion that the voltaic Chain of Mr. Pulvermacher, while it presents an instrument of the requisite efficiency, has the advantage, for medical purposes, of being more portable, more readily available, and more economical in its use, than the electro-galvanic instruments at present employed.*"

Letter from VALENTINE MOTT, M.D., LL.D., *Emeritus Professor of Surgery and Surgical Anatomy to the Faculty of Medicine at New York, Honorary Fellow of Queen's College at Dublin, Ireland, etc., etc.*

"DEAR SIR,—I have been much pleased with the Hydro-Electric Chains of Pulvermacher, which you have been so polite as to furnish me. They are a very ingenious and beautiful arrangement of the galvanic principle, and I have no doubt will lead many medical practitioners to use this powerful agent, from their neatness and convenience. They are so portable, and at the same time powerful, that many will resort to them, and indeed be pleased by them, who would be alarmed at a more complicated apparatus.

"*I think it may become an important remedial agent, and*

its efficacy may be greatly increased if used in the manner of galvano-puncture.

"You have my best wishes for its successful introduction.

"Yours respectfully, "V. Mott."

"I have employed Pulvermacher's Hydro-Electric Chains, and am satisfied *that they constitute the most effective and uniformly available apparatus* for the therapeutic application of electro-galvanism of which I have any knowledge.

"Wm. H. Van Buren, M.D.,

"Surgeon to the Bellevue and St. Vincent's Hospitals, Professor of Anatomy in the University of New York, etc."

"I have made trial of Pulvermacher's Hydro-Electric Chains, *and regard them as a very convenient and efficient contrivance for the application of electricity in the treatment of disease.* The portability of the instrument and the facility of its application are such as to render it superior, in my opinion, to any other electrical apparatus employed for medical purposes.

"Alfred C. Post, M.D.,

"Professor of Surgery in the University of New York."

PATENTS AND PIRATES.

From "The Stock Exchange Review," Dec. 1874.

The conferences and consultations of men of business and science on the Patent Laws, and deputations to Ministers of State lately reported in the newspapers, had cogent cause and useful definite objects. The general result encourages the hope of a general amendment of the system.

If the Patent Laws are to be retained at all, it will be specially important to make efficient provision for the repression of a nefarious class who to some extent bring discredit on the most valuable discoveries by foisting counterfeits on the public, and by the equally fraudulent device of imitating the advertisements and descriptive pamphlets issued by an original inventor, and even attempting to create a false idea of actual identity with him.

Probably in no branch of science bearing on the conservation of health, cure of disease and pain, and prolongation of life, have knowledge and practical efficiency made such marked advances during this last quarter of a century as in the now extensive domain of Medical Electricity. . . .

An interesting instance of its efficacy is remembered by some of the old school of literary, theatrical, and other persons connected with metropolitan amusements. A few years ago, one of the best known and most successful caterers for popular enjoyment, who was long the victim of neuralgia, rheumatism, and cognate miseries, was about celebrating his birthday with the customary genial festivities. Early in the day the house was alive with "troops of friends," but the hero of the happy event lay a prey to the horror of acute face-ache, and totally unfit to greet or meet his guests. Suddenly it occurred to one of his daughters that in one of the cupboards was a Pulvermacher Chain-band, which had given prompt relief some months previously, when papa was suffering from an attack of sciatica in the hip. Perhaps it might do good now? No sooner thought of than tried. The Chain was applied over the temple and down to the chin, "and in twenty minutes," to quote the exact words of the host, "I was one of the merriest fellows in the company."

. . . The most eminent medical authorities have certified their appreciation of the important services rendered by Mr. Pulvermacher in having brought Medico-Electrical science to its almost perfect condition.

Yet it is this peculiarly meritorious worker in the field of therapeutic science, whose labours strike especially against pain, disease, and premature death, who is specially the object of annoyance from the fraternity of pirates, quacks, copyists, &c. They have pirated the headings to his descriptions, they have pirated the descriptions themselves, they have pirated (verbatim) his very testimonials, in order to make it be believed it was the counterfeit base wares the eminent writers had approved and praised. . .

Mr. Pulvermacher has obtained injunctions against some of them, accompanied by indignant judicial denunciations of their misdeeds. But they are a difficult brood to deal with. Their number is legion, their aliases could not be counted without drawing breath, and when necessary they can be ubiquitous as to their local habitations.

MEDICAL ELECTRICITY—ITS MERCENARY ABUSE—USEFUL REVELATIONS AND HINTS.

THE age has arrived at such a condition of scientific knowledge that, when a glaring fallacy or mis-statement is made about matters connected with the positive sciences, such as physics or chemistry, its true value can be easily shown by experiment, and so conclusively, that any person who affects to make light of the proof simply shows his ignorance and his contempt for Nature and her unchangeable laws.

To take a very homely illustration, let us suppose that some wiseacre recommended us to use slate instead of coal, and assured us, that by so doing, we should save fuel, prevent the nuisance of smoke, and obtain more heat. Few people, if any, would think seriously of such a piece of advice, but would set down the adviser as either an ignoramus or a fool.

Now, a statement equally idiotic, though considerably more specious, has been made on the subject of Electro-Therapeutics, to the effect that certain curative agents (which have been largely advertised) depend for their alleged curative power upon Magnetism, for which is claimed attributes possessed exclusively by Electricity, and indeed, said to be almost identical with it. Others have put together pieces of zinc and copper in a haphazard manner, and asserted the same to be Electric; nay, more—this admirable curative agent is set forth by adventurers as being contained even in the perforated plaster so largely advertised; but men of real scientific knowledge will at once mark the inaccuracy of these statements, and rate them at their true value, so that no harm can accrue to science. It is, however, difficult for the great bulk of persons to distinguish between the real and the false on these subjects, because Electricity and Magnetism form a comparatively recent branch of knowledge amongst us, and the world in general has not had time to become familiar with them. Few people, indeed, are aware of the immense amount of literature that now exists connected with Medical Electricity, proving how very important a part it plays in the treatment of disease. No curriculum or scheme of medical education at any of our English hospitals would be considered complete if it did not include Electro-Therapeutics; while on the Continent, and particularly in the Medical Schools of Paris, much encouragement, in the shape of valuable prizes, is given to all who devote themselves to researches in this direction. From the slender acquaintance of the mass of the people with electrical knowledge, arises a great evil, viz., that the widest possible scope is afforded for unscrupulous speculators who think that they can say pretty much what they like, however untrue or ridiculous, without fear of exposure, and identify themselves so cleverly with the inventions of those who have really studied the subject, that it is very difficult to distinguish the wheat from the tares, or to know how much to believe and how much to reject. It has ever been the misfortune of true science to be infested with this class of parasites, and Electricity, for obvious reasons, has suffered more than others. It has happened more than once that some of these speculators, seeing the great success which has always attended the true Voltaic appliance, composed of zinc and copper, arranged scientifically, have brought out spurious articles, also containing zinc and copper; but as these imitators are completely ignorant of the elements of electric science, it is scarcely to be wondered at that their devices, whether charged with acid or not, are perfectly powerless to produce Electricity.

At the present day the attack made against Electricity comes in the form of Magnetism, which, though discovered long before it, has never yet been proved to have any curative powers, although at different times enthusiasts and interested adventurers have tried to make us believe so. If we search the archives of scientific medical literature, we find nothing to establish Magnetism as a remedial agent, while scarce a single authority of any note has ever decided in its favour. This wonderful idea has been left to be discovered, curiously enough, by the vendors of the extraordinary material termed "Skeuasma," and the still more extraordinary "Magnetine." They, knowing that people in general are willing to accept a statement as granted, ask us to believe that Magnetism has now for the first time taken its place side by side with Voltaic Electricity, the reputation of which they have borrowed, in the hopes of confusing the one with the other in the minds of inquiring patients. More curiously still, they base the connection between the two on the fact that magnets are used in the magneto-electric Faradaic machines, but they calmly ignore, that *mechanical force* is indispensable in order to convert thereby movement into Electricity, as is also the case in the old frictional electrical machine; whereas the Voltaic Battery *yields it without any motive power,* but simply by chemical reaction.

Frictional Electricity—previous to the discovery of Galvanic Electricity—had been used with marked success by, amongst others, the celebrated John Wesley;* but as soon as the identity of the two was scientifically proved, and the causes of the different mode of action clearly shown, it became at once evident that Galvanic Electricity, as a curative, would prove a much greater success. For the only difference is that whereas *Frictional Electricity* is devoid of chemical effects, is of excessive tension, violent action, and very unmanageable to the patient, *Galvanic Electricity,* on the contrary, is analogous to the vital functions, gentle in its action, and in its working congenial to the nerve power in the animal economy. It is, moreover, most simple, practical, and effectual in application, and is so vastly superior in healing power, that it has entirely superseded frictional electricity, as the galvanic battery for telegraphy has superseded the optical telegraph or semaphore.

But the would-be *savans* above referred to, having no idea of the nature of this identity, ignorantly claim it for Magnetism, and speak of "Magnetic currents" as though such existed, whereas we all know that

* "The Desideratum; or, Electricity made Plain and Useful." By John Wesley. Republished by Baillière, Tindall, & Cox, London, 1871.

Magnetism is simply a *static* force, whilst Electricity is either *static* or *dynamic*, at will.

The material in which these so-called magnetic contrivances are enclosed being flannel, it might be a reason for the projectors to claim relationship with Electricity, because it is often used by electricians for rubbing a glass rod to demonstrate the principle on which the frictional electrifying machine is based. But flannel being considered by them far too ordinary and vulgar for inventive genius of such a high order to associate itself with, a mysterious compound is introduced by the insertion of pieces of india-rubber intermingled with steel filings, thus throwing dust into the eyes of the would-be believer. Then, and then only is it pretended that the curative powers assert themselves, the great inducement to the purchaser being that the garments, in which these miraculous effects are hidden, are so easily put on and worn. To me it seems that the garments, as garments, would be better without these little arrangements; and that just as much, if not more, good would be obtained by the wearer from the friction of simple flannel than from the intricate contrivances contained in it. I may, however, safely assert this much, that I do not for a moment believe that the latter can do any harm; so that if people have a fancy for buying expensive under-garments curiously inlaid with india-rubber and iron filings, or little steel plates, there is no reason why they should not gratify it.

As far as their real scientific value is concerned, we are fortunately able to gauge it pretty accurately, knowing, as we do, that physical agencies are regulated and controlled by positive natural laws, and this, indeed, is the reason why we are better acquainted with the *modus operandi* of Electricity than we are with most of the remedies of the Pharmacopœia. Hence the vendor of a quack medicine has the advantage over a physical remedy. He makes a pill, for instance, advertises largely, and induces the public to swallow it, because it believes that there is some new drug of wonderful qualities lying hidden in the compound, which is the cause of the health improvement promised in the advertisement. But a natural remedy like Electricity does not admit of any of this convenient secrecy or mystery, for everybody can try it for himself, and its powers to cure are demonstrable by physical and physiological experiment.

The whole system upon which this so-called magnetic cure has been got up proves that it was intended to graft it upon the reputation of Electricity, and, in particular, upon those electrical appliances which have been the study of my whole life. In the first prospectus issued by these imitators, they started by boldly adopting my motto, "Electricity is Life;" but finding that I had obtained an injunction in Chancery against an unscrupulous plagiarist, they speedily dropped the motto, and took instead a graphic picture of a magnet staff, illustrating the magnetic force by the curve lines which result from scattering iron filings on a card board under which a real magnet is placed. But if they were asked to produce this phenomenon through the agency of their absurd india-rubber magnets, they would be unable to do so. Before it obtained its new name "Magnetine,"* "Skeuasma" had a very poor time of it, and lingered on obscurely and ingloriously, until a coadjutor of these *quasi*-magnetists addressed the following letter to the Editor of the *Christian World*, in which he called attention in the most philanthropic way to his appliance, and suggested that it was worthy of standing side by side with mine :—

"To the Editor of the "Christian World.'

"SIR,—Not long since there were two communications inserted in your columns of a very interesting character, in reference to the curative properties of Pulvermacher's Galvanic Chains, quoting cases of their successful application so remarkable that many of your readers would probably be rather incredulous as to their accuracy. For myself, I had no hesitation in accepting them, on two grounds—viz., that from a long personal acquaintance with both of the writers, I am satisfied that the cases are not likely to be mis-stated, or even over-stated ; and secondly, that I can, from actual knowledge, confirm the truth of one of the instances given.

"My present object, however, is, with your kind permission, to call attention to another patented invention, somewhat similar in principle, purpose, and practical use, worthy, at least, of standing side by side with Pulvermacher's, but not at present so extensively known, although already advertised in the *Christian World*. I mean 'Darlow's Patent Magnetic Belts' and other appliances. My own interest in the matter was excited some few months since by meeting quite accidentally with a remarkable case of recovery from Bronchitis by the use of these applications, after the patient had been quite given up by the doctor ; which induced me to persuade a friend who had been a great sufferer from the same complaint (not having enjoyed a comfortable night's rest for six years past) to make trial of the same remedy.

* **Their having now changed the name Magnetine into Ferro-Magnetine is highly suggestive. If there is *no Magnet-in* there is avowedly at least *Iron-in*.**

He found immediate relief, and within a week or two, all the distressing symptoms had completely disappeared. Since then several other equally remarkable cases have come under my notice; and last, not least, by the advice of Mr. Darlow, I tried the effect of a Belt for Bilious Sick Headache, with which I have been troubled for more than forty years. A slight attack occurred some three or four days after I had worn the Belt, before, as I presume, it had fairly begun to operate upon the system; but since then (now between three and four months) I have been entirely free from it, and in addition, have realised a very remarkable degree of elasticity and vigour. In the interest, therefore, of suffering humanity, I venture to request the insertion of this, which would at the same time be regarded as a personal favour by

"AN OLD SUBSCRIBER.

"Stoke Newington, February 28, 1871.'

My reply was as follows:—

"A DISCLAIMER.

"*To the Editor of the 'Christian World.'*

"SIR,—By the insertion of a letter in the *Christian World* of March 10, 1871, in which Darlow's so-called Magnetic Skeuasma is placed side by side with Electric Chain Bands, &c., a confusion has been created in the minds of the readers of your paper—a confusion which, in justice to the public and in the interest of the real progress of Medical Electricity, it is most important to clear away. If mineral magnetism were a curative agent, and as such adopted (an hypothesis by no means warranted by experience), still the Skeuasma could not reasonably claim to contain magnets or electricity, and the fact may be demonstrated by a very simple test. A small magnet (to be purchased for a penny in any toy-shop) at least attracts soft iron filings, while the pieces of india-rubber placed between the flannels of the Skeuasma do *not*; consequently any ordinary flannel chest-protector will have the same effect as the Skeuasma.—Your obedient servant, "J. L. PULVERMACHER.

"200 Regent Street, W., May 23, 1871."

To this they made the following reply:—

"It was to be expected that Mr. Darlow and his friends

would resent the allegation contained in the note of Mr. Pulvermacher, inserted in these columns May 26, that an ordinary flannel chest-protector would have the same effect as the one he has to sell; and it is only fair for us to let it be known that several correspondents, including Mr Darlow himself, challenge the fullest investigation of the merits of the invention of which Mr. Pulvermacher speaks with so much contempt.

"Mr. Darlow writes:—'My attention had been called by numerous letters to one which appeared in your columns May 26, entitled "A Disclaimer," in which the writer asserts that the Magnetic Skeuasma strips are not magnetic *because they will not attract iron filings.* Such a conclusion from such a test is erroneous and unscientific—the *true test* of a magnetic substance being that it shall exert *both* an attractive and a repellant influence on another magnet, such as a common pocket-compass. This test can be as easily applied as the one quoted by the writer of the said letter, with the advantage of being simply unanswerable and scientifically sound. The Skeuasma answers to this test powerfully, even though several inches of solid matter be interposed between a bundle of the magnetic strips and the needle. It also retains indefinitely the magnetic influence, unless tampered with, in a superior degree to all other magnetic bodies, and, in virtue of its magnetic properties, it becomes, when applied to the body, a constant source of electric power. As to its curative value, we are content to let the public and their testimonials speak for themselves.'

"Our previous correspondent, 'An Old Subscriber,' strongly re-affirms his former testimony to 'the curative powers of Darlow's Magnetic Appliances,' and states that he has had considerable additional observation of their most beneficial use. 'In some cases,' he says, 'the results have been marvellously rapid.'"

This argument was so essentially childish, that I really did not think it worth answering; for everybody knows that this attracting and repelling influence is exerted equally well by a freshly-sharpened knife. A needle, if exposed to terrestial magnetism, with which we are constantly surrounded, will exercise as great or more influence on the needle of the pocket-compass than a ton of the so-called Magnetine.

This test as a proof of efficacy is a delusion, for the same can be done with a common door-key—the fact being that the magnetism is in the needle of the pocket-compass employed, and is of course attracted by the iron-filings contained in the so-called "flexible magnets"!! yclept Magnetine.

This experiment is evidently in imitation of that made with an electric current on the needle of a galvanometer,* although this remarkable difference exists, viz., that in the latter case the electric current conveys its magnetic influence to a needle at long distances, as across the Atlantic by cable.

To make it appear as if they had gained the palm in the controversy alluded to, *my silence was made a great point of* by these magnetine promoters in their pamphlet; and as they went on unmolested, they became bolder, and represented Galvanism as injurious to health, and quite superseded by Magnetism, although a little before this they were glad enough to claim for the latter the same curative effects as those produced from Galvanism, thus attempting to knock down the ladder by which alone they had achieved the little notoriety they possessed.

These self-styled "Professors of Magnetism" talk very largely of the bad effects of shocks and strong corrosive acids, commonly used in the old batteries, but they take good care to ignore the fact that the

* See page 27.

success of my Chain-Bands is entirely owing to my having dispensed with shocks and corrosive acids, and substituted simply a little ordinary vinegar and water. It was this special advantage that was so highly praised by Dr. Oppolzer, the celebrated Clinical Professor at Vienna, in 1849. He also honoured me with a testimonial, which I still have by me, and which, with others of the same class, show that the Volta-Electric Chains had a recognised position before any list of private cures effected by them had been published. (See Extract of Report of French Academy, page 32.)

To patients who think for themselves, and are not led away by clap-trap and unscrupulous statements, one testimonial from such an authority will out-weigh thousands of others from private sources. The reader of the advertisements of these magnetarians may well be confounded, when he knows that Magnetism cannot possibly bear comparison with Electricity as a curative agent; because if Magnetism had any remedial effects, there have been plenty of investigations and sufficient time to enlighten us as to its efficacy; and had it stood the test as Galvanism has done, it would, in like manner, stand on its own merits.

Magnetism cannot be used for producing Electricity in a garment, it being impossible to introduce all the mechanical combinations of the Faradaic electro-magnetic machine in a portable form. Portable Magnetism has been tried over and over again without success; and I may mention that, as long ago as 1849, a Patent (No. 12,847 *Old Law*) was taken out by Mr. Meinig, who afterwards abandoned it for Voltaic Electricity.

The reader may be curious to know why portable

magnetic appliances have failed. The following reasons are familiar to every student of physics :—

1stly. Magnetism, in a *direct* application on the body, even were it a curative, would have to be applied with great power, exceeding that exercised by the terrestrial magnetism, with which we are everywhere and constantly surrounded. If a weak influence had curative effects, it is amply supplied by the terrestrial magnetism, thus rendering superfluous any artificial appliance.

2ndly. The degree of permanent magnetic power depends on the weight of the steel employed; therefore a really magnetic portable appliance is an impossibility, owing to the bulk and weight of the magnet or magnets which it would entail.

3rdly. The magnetic effect being circumscribed, and fixed on the poles of the magnets, is confined to a radiant action, and is not conductible, but static. Its power diminishes by distance in geometrical proportion; thus, the strongest wearable magnets would be reduced in effect to an infinitesimal degree, if separated from the body by any layer of, say india-rubber or flannel.

4thly. Magnetised steel loses its magnetism by heat, vibration, and oxidation, so that even the warmth of the body is sufficient to disperse it.

Moreover, Magnetism, had it a specific curative action of its own, is nevertheless a superfluity, its effects being embodied in those of Electricity, which converts every conductor through which it passes into a magnet.

The idea set forth by the vendors of magnetic contrivances, that the iron in the blood, circulating before the poles of a magnet, becomes electrified, as in the magnetic machine, is a gross fallacy; since the iron cores in the latter can only come into action when surrounded by coils of isolated conducting wires, and made to *move* within the sphere of influence of the poles of strong magnets, so as to induce the electric current therein. It is a piece of outrageous absurdity to assert that the same thing happens to the blood, which is not in this condition, or even in the necessary metallic state for entering into any such physical relation. These absurd doctrines are calculated to confuse the public mind in discriminating between the false and genuine appliances, and are so flagrantly fallacious that, unfortunately, none of the many scientific men will condescend to contradict them.

There is no fact extant, nor any scientific theory existing, which could warrant such assertion; unless reference be made to a *quasi*-scientific work propounding a *fanciful theory*, concocted to order by some would-be *savant*, such as is mentioned by Molière in the character of "Sganarelle" in "Le Médecin malgré lui," who, when catechised as to his error in regard to the relative positions of the heart and liver, exclaimed, "*Nous avons changé tout cela.*"

But what shall we say of those two or three medical men who apparently countenance such contrivances? Simply, that their want of knowledge on the science of physics is lamentable. Any scientific man naturally finds it an ungrateful task to contradict statements so utterly puerile and unwarranted, but he may be compelled to do so in the interest of sufferers, many of whom are unable to discriminate between the real and the false. Consequently, for them to be forewarned is to be forearmed.

If a *bonâ fide* magnetic appliance could be found, of any real curative powers, no one would be more ready to give it a fair trial than I. *My Patent Safety Pin Fastener, being also made of magnetised steel, and studded on under-garments, is a most convenient means of application, and is at the disposal of any one who would try it fairly and report faithfully on the result.** This application of portable magnets to the body, has the advantage of not requiring any fastening by means of tapes, being easily attached on the under garment, and detached when the latter requires washing. Or if a Magnetic Belt be required, the *insulating wrapper,*† which ensures the warm and dry application of my Galvanic Chain Bands, &c., can be easily metamorphosed, by simply fastening inside it a dozen of the magnetised pin fasteners (price 2d. each, or 1s. 6d. per doz.). As there are always persons to be found who will countenance competitive innovations, that are opposed to a well-established and successful invention, Mr. Pulvermacher has arranged to keep on hand a good supply of these Safety Fasteners, made of magnetised steel; so that those who may have a strong craving for magnetic influences, may be able to gratify and test their notions at little or no expense to themselves.

The looker-on may well feel not only amazement, but amusement, at the persevering and systematic manner in which these medico-magnetists persist in forcing their wares on credulous people. As far as the injury to the purchaser's pocket goes, we may be amused at the cleverness by which he is persuaded to make an outlay for this wonderful hosiery. But there is a more serious side to the question—viz., the disappointed hopes which are felt by the sufferers, who, only partially under-

* See full details in the Prospectus relating to this invention.

† See page 40.

standing, but wholly believing, the advertisements, don the magic garments in the expectation that all their pains and aches will disappear, believing that they are about to try Medical Electricity in an improved form; and are thus diverted from trying the only *bonâ fide* portable electrical appliances in existence, the invention and perfection of which have cost a lifetime of study and labour.

The rampant spirit of competition is, however, not confined to would-be inventors: it crops up in a variety of forms. As an illustration, a business establishment is opened by certain "doctors," ostensibly for the sale of appliances of the above description, but in reality with other ulterior objects. Publications are got up as a guide, for drawing the patient to their place of business, and made to appear by colourable imitations as in plausible competition with mine. A sufferer, when applying for one of their advertised contrivances, is under the impression that he can use it himself without medical assistance; but being persuaded by the manager's grandiloquence that the serious and complicated nature of his case requires medical aid, he is induced to place himself under the treatment of the "medical electrician," the real proprietor of the establishment, who, once having him in his grasp, subjects him to all kinds of extortion.

Again, some shady members of the medical profession, after an unsuccessful start in ordinary practice, suddenly become "medical electricians," without any qualification whatever; being ignorant even of the indispensable rudiments of that branch of physical science which they pretend to apply for the cure of disease. This, however, is of little consequence to them, provided they can assume the character of a learned philosopher in Electricity, so as to be able to secure the confidence of patients. To obtain a claim to this character, and push themselves to the front, they have recourse to professional book-writers, always to be found hovering about the British Museum; and engage one to compile a nice little volume from the interesting facts disseminated in the works of great authors on Electro-Physiology and Electro-Therapeutics. This, of course, is made to appear to the uninitiated as a highly creditable production of the would-be electrician. Improvised in this fashion as a scientific character, and thus raised from his native obscurity, he, from the moment his name appears on the cover of the treatise concocted for him, assumes the bearing of a high authority; as though he himself had discovered all the scientific facts, and invented all the ingenious instruments, which the compiler of his book has found demonstrated in the works of gifted and laborious investigators. He finds that appliances in Medical Electricity, which admit of being applied by the patient himself, do not suit his interests; therefore he naturally "runs down" that which stands in his way, although it has stood the test before the public for many years. Knowing that any cases of successful treatment, which he might bring forward as his own experience, could not be substantiated, he adopts the prudent course of compiling them from other sources; and herein is one of the criterions by which his pretensions may be detected, and his deficient knowledge of Electricity and Electro-Therapeutics made apparent.

If, by the foregoing observations, I have been able to induce any of those in search of a genuine remedy to pause before they commit themselves to the tender (?) mercies of any of the worthies referred to, I shall not have written these lines in vain.

www.ingramcontent.com/pod-product-compliance
Lightning Source LLC
Chambersburg PA
CBHW021943190326
41519CB00009B/1125

* 9 7 8 3 3 3 7 8 6 4 2 1 7 *